EFFECTIVE STRUCTURED PROGRAMMING

EFFECTIVE
STRUCTURED
PROGRAMMING

Lem O. Ejiogu

PBI
a petrocelli book
new york / princeton

Designed by Diane L. Backes
Typesetting by Backes Graphics

Printed in the United States of America
1 2 3 4 5 6 7 8 9 10

Library of Congress Cataloging in Publication Data

Ejiogu, Lem O.
    Effective structured programming.

    "A Petrocelli book."
    Bibliography: p.
    Includes index.
    1. Structured programming. I. Title.
QA76.6.E425   1983   001.64'2   83-2206
ISBN 0-89433-205-8

To
all my teachers
here
and
everywhere

CONTENTS

PART TWO
Metrics of Software Engineering

———————◆———————

PREFACE

Structured programming is structured thinking.

Above all, structured thinking is a rational, logical organization. It transcends any art. It is the result of cumulative, learned, reliable behavior. Its well-defined principles are based on a set of axioms, implicit or explicit, and are capable of being learned or verified by any prepared initiate. Its purposes or products follow rationally from the direct application of these principles.

The philosophy of structured programming seeks to eliminate tricks from programming, thus producing provable, reliable software systems. Its purpose is to raise programming from the status of art to one of applied science and to erase myths surrounding it.

However, since the advent of structured programming, there have been diverse opinions and even confusion about what constitutes structured programming [10, 14, 16, 31, 35, 36]. Among the interpretations are structured coding, structured flowcharting, a DOWHILE-DOUNTIL, IF-THEN-ELSE construct, a top-down approach, a chief programmer team management, a structured walk-through process [22], and pseudo-coding and documentation [8, 47].

It is the purpose of this book to search through this jungle of perceptions and to identify the main principles for applications. In doing so, I have presented structured programming as a tetrad of applied philosophy. These four faces are:

Managerial (chapter 2)
Aesthetics (chapter 3)
Benefits (chapter 4)
Principles (chapter 5)

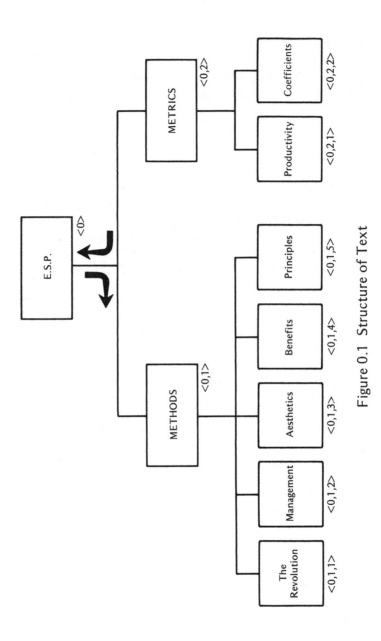

Figure 0.1 Structure of Text

On the surface, my observations lead to the following conclusions:

1. The techniques of chapter 2 are sufficient but not necessary.
2. Those of chapter 3 are necessary but not sufficient.
3. Those of chapter 5 are both necessary and sufficient.
4. Chapter 4 is a reaffirmation that the goals of structured programming necessarily and sufficiently justify the practice.

Finally, I have constructed a set of simple metrics on (structured) software. The values obtained using basic arithmetic reasoning far supersede any I know in use today. These are the contents of Part Two (Figure 0.2). Chapter 6 offers a general formula for programmer productivity and a special formula for productivity measures, and uses some of these coefficients to obtain an *effective* value of a program.

Chapter 7 develops the determinant factors of degree of complexity, degree of structuredness, degree of requirements definition, degree of correctness and degree of reliability of software. In principle and practice, managers now have a concrete tool for obtaining a numeric measure of a programmer's productivity and for evaluating productivity planning and software acquisitions.

Throughout the book, particularly in arithmetic expressions, I have refrained from rigor so as to make the book readable for professionals. The professional teaching faculty will find it easy to comprehend. Criticism is encouraged to resolve the confusion about what constitutes structured programming.

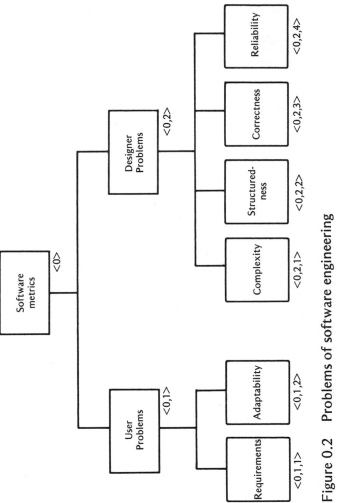

Figure 0.2   Problems of software engineering

ACKNOWLEDGEMENTS

I am timelessly thankful to the following without whose moral and material support this manuscript could not have been written: Christopher N. Chijide, Godwin O. Edionweme, Obed O. Okechukwu, Jonah Rosenberg, and the Lemieux family—Marc, Gisele (whose reverence for living has constantly inspired me since my college days), Louise and the late Pierre (in memory).

With gratitude, I acknowledge the permission to adapt figs. 3 and 4 of Böhm and Jacopini's "Flow Diagrams, Turing Machines and Languages With Only Two Formation Rules", *Communications of the ACM*. For their kind services, I thank: the Science & Technology Division of the Chicago Public Library, the Kemper Library of the Illinois Institute of Technology, the University of Illinois Circle Campus, and Mrs. Janice C. Fox of ICCP, all in Chicago. For their encouragement, I thank my mentors—Eugene A. Cordury, Eric K. Ericson, all Snr. geologists at Gulf Oil Corporation (1960-66), and the late Rev. Charles Conder of San Diego; and also, Isaac Aamidor, Edwin Okabuonye, Clifford Rowland; and my respected aunts—Mrs. Joanah Iroka and Mrs. Charity Onyewuchi.

On the professional front, I am grateful to: Lee Stone, Ted Rock, Terry Schroeder, Robert Kurek, Jim Jagusch, Frank Zilic, Jim Macafe—all of Official Airline Guide; Wayne Oberschelp, Richard Probst, Jim Collins, Jim Holder—all of American Hospital Supply Corporation; and Girish Parikh, Independent Consultant, whose curiosity fortified my resolve to put these stimulating ideas into writing.

Finally, I must especially thank: Len Neuzil, formerly of Montgomery Ward & Co., whose inspiration propelled me to new vistas; Dr. S.J. Bryant of San Diego State University, whose provocative definition of abstraction has become my guiding light; and Dr. J.D. Kinloch of East Tennessee State University who taught me to rigorously strengthen the strong points of my argument.

PART ONE:

METHODS of SOFTWARE ENGINEERING

"For we know in part, and we prophesy in part. But when that which is perfect is come, then that which is in part shall be done away."

St. Paul's First Letter to
The Corinthians; 13:9-10
(King James version).

Chapter 1

THE REVOLUTION in PROGRAMMING

Evolution is the fundamental principle of all nature. In the physical world, it manifests as observable growth; in the intellectual world, as progressive learning; in the mental world, as cumulative wisdom; and in the spiritual world, as systematic creativity. Nature evolves from the simple to the complex, from the amoeba through to organs and animals to man, but the human mind comprehends the complexities of nature through their differentiated simplicities.

1.1 "SPAGHETTI" SYSTEMS

Great advances have been made in software engineering, but not without some frustrations and sweat on the part of pioneers. Pioneers are always instrumental to the creation of great ideas; however, the very nature of evolution dictates that certain methods be, at some time, supplanted by other measurably effective ones. This observation has been substantiated in the design and programming of computer systems.

Traditional methodologies for developing computer software systems revolved heavily around flowcharting. Though not without some accomplishments, flowcharting has become a double-edged sword. On one edge, it looks simple and innocent and even as attractive as an engineering blueprint; but on the other edge, the effort, time, cost and effects of using it to develop software has been overwhelmingly expensive and disastrous. Such systems have inherent in them the disquieting attributes of unmanageability, unmaintainability, unmodifiability, unverifiability and many more. For large systems, the flowcharts become very bulky, discontinuous (scattered over several pages) and incoherent. They characteristically lack dimension. Hence, the use of the expression "spaghetti" systems.

3

Flowcharts begin somewhere; then branch left and right, up and down, from page to page, back again; then off again, repetitiously, on and on. There is a general loss of control. The entry and exit points are almost unrecognizable. For large systems the maze is almost endless. Trying to trace the flow of a given data name is like looking for a needle in a haystack. Trying to trace backwards for the source of a data item is like labelling the sources of drops of water at a river confluence.

Perhaps prompted by machine language, the thrust toward flow-charting was the development of such methodologies as data-driven, information-hiding techniques. Data or information in these techniques was seen as discrete entities whose passage through the hardware system was representable as graphlike pieces of usable commodities. With these techniques, the concept of functions was never emphasized.

But a more powerful potential enemy of structured thinking has been the uncontrollable JUMP statement called GOTO. GOTO is found in several of the programming languages, and, theoretically, has its roots in flowcharting. The flowcharts had devices, such as connector circles for jumping from one page to any of several pages as well as conditional and unconditional jumps. It has been admitted that this discontinuity of thought has been very damaging and costly to development, design and maintenance of computer software.

Another concept that baffled the data processing industry was modularity. Its many facets of definition and measure were engrossed in confusion. Some people postulated a fixed number of lines of code; some, arbitrary input/outout size, and others, a single function. Some designers would even call an ordinary algorithmic construct a module; there seem to be no formal theorems to draw upon. Flowcharting was part of the development of these modules, but was fortunately discouraged by Parnas [41]. Recently a formal theory of modularity has begun to emerge, postulated along the theories of structured programming [39, p. 94]. However, the author believes that a module is a mathematical system—strictly a semigroup under the intersection operator.

The author is not suggesting programming along the methods of mathematical reasoning (see Figure 1.1A), but merely concurs with Dijkstra that in spite of all its rigor and approximate exactitude, mathematical reasoning offers an outstanding model of how, with a brain of limited capacity, to grasp and simplify extremely complicated structures [18]. Such a mathematical knowledge can be extremely useful for designing software systems. This is the value of a theory. If we know the nature of a thing, then we are better prepared to use that thing with greater confidence. It is lack of this confidence that has generated the crisis and therefore the present revolution in software engineering.

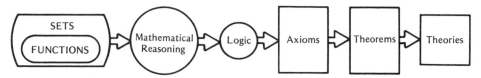

Figure 1.1A   Mathematical reasoning—a tool for obtaining results from observations (of resources) through abstraction

Another technique that was less completely mastered—probably because of their cumbersome size, especially for a large number of variables —is Decision tables. These are essentially a generalization of the truth tables of symbolic logic, now studied as switching theory. Some people advocate their increased application [28].

Another cause for the setback in software evolution is the proliferation of higher level languages. Some had thought that it was both necessary and possible to invent some "ideal" programming language that would be a panacea for all programming ills. Time has proved such a speculation false, for each language has some characteristic limitations and inadequacies. In practice, certain programming languages are more powerful than others for specific software applications. What is needed is a simultaneous evolution of these languages through incorporation of some of the useful aspects of structured principles (chapter 5). Some language designers are already doing this. For example, many of the recent amendments being made to FORTRAN could well make it a reliable, powerful programming language.

1.1.1 TRADITIONAL DESIGN METHODOLOGIES

This section briefly outlines some of the older techniques for designing systems and programs. The concept of function is nearly totally absent in its literature. It is remarkable that these methodologies have contributed in some measure to the search that has culminated in the structured revolution.

Procedure-Centered Design

This is perhaps the oldest approach to systems design; its philosophy is built around the procedural representations of activities. Consequently, the flowchart is at the heart of its application. Any module can call another, depending on the logic desired. The decisions about which modules belong to which load units are primarily the designer's to make, using such secondary decisions as size, frequency of call to a module, or presence of a SORT in a module.

Data-Driven Design

This is the antithesis of the functional approach to design. It assumes, perhaps incorrectly, that data is explicitly defined, unlike function, at the beginning of systems. Thus, the emphasis is to [13]:

1. Identify clearly inputs and outputs.
2. Determine the data structure in (1).
3. Define the datapath in terms of statement sequence, controlled iteration, and explicit selection.

For structuring data, some of the practicing conventions utilize tree diagrams, or Warnier diagrams. Data dictionary organization is formulated after activity (1) is concluded. This methodology can lead to static or unmodifiable systems. In addition, it may be argued that these methodologies are not usable for design of experimental systems where the outputs are unknown.

Information-Hiding Design

This technique requires certain data (stacks, pointers, arrays, etc.) to be "concealed" within some modules. These concealed data are made explicit for some modules, and these modules are then bound together. It is clear from this that it is data-control oriented. We will illustrate this approach with an example.

EXAMPLE: 1.1.1

The function computed by

WORK = IMPEDANCE * EXP (SINE (THETA))

does not declare if the EXP function is a subfunction, subroutine or subprogram, or if SINE or THETA is a vector (array) or a scaler. The same can be said of IMPEDANCE: Is it a constant, a sub-function or something else? These issues are resolved by the particular modules that define these data structures. This technique is also called Parnas's technique [41].

Transform-Centered Design

This technique is globally oriented, i.e., it abstracts those activities that are related and factors them into the subgoals of operation. These then become the "modules" of the system. The purpose is to isolate the primary processing activities according to their order of (data) flow. This leads to the concept of flow diagrams with input data and output data connected by arrows to the point of processing activity. This can lead to circuitous diagrams—giving rise to spaghetti systems.

Transaction-Centered Design

As the name suggests, the types of transaction present determine the types and numbers of modules in this systems design methodology. At the top of the system is a general module that receives, differentiates and passes transactions (records) to their corresponding modules, essentially "splitting" them according to their known destinations. From here on, any of the previous methodologies may be called in to process the transactions.

Data-Structure Design

This methodology (attributed to Michael Jackson and Jean D. Warnier) [53, p. 223] builds program structures based on the data structures. It is clearly insensitive to the actual functions implicitly or explicitly implied by the system. Its greatest disadvantage is structure clash [29, p. 151], where more than one set of data having different data structures is to be described for process. There is inconsistency in its program structure.

1.1.2 THE FRAILTIES OF FLOWCHARTING

The most powerful incentive to program flowcharting is the SELECT structure, in particular, the nested-IFs. Flowcharts emphasize program flowpaths instead of program logic—a case of putting the cart before the horse. Program logic almost exclusively determines the output data a user expects from a program. Even for programmers, flowcharts are hard to read.

Essentially, flowcharting is a mechanism for refinement; however, we shall show here many drawbacks of flowcharting, including those that are euphemistically called structured flowcharts, a misapplication through misinterpretation of the structured revolution. We shall also demonstrate that these traditional program flowchartings have *absolutely* no place in structured programming, from analysis through design and realization. Instead, we need to use much more simplified structures that are well-known and consistent with our "structured thinking." We shall show that the nested-IF structure is a *binary tree*, a well-studied and documented structure in the literature of information systems. We simply "descend" from the root node down the tree to access any data or function.

The critic may argue that the tree diagram is a type of flowcharting. Yes, but, some types are logically and practically better, more informative and more manageable than others. If types exist, they may always be grouped into classes and subclasses to enable comparison. It is this ability to compare that leads to better judgment in the choice of strategies.

Figure 1.1.2A is a traditional flowchart diagram of a simple interface (among three communicating functions) of a program. Three arrows are entering at A, but there is no way to tell the source or destination of a specific data element. Some data are coming from 11 or 12, or 13.

Figure 1.1.2B illustrates a conventional practice in "structured" systems analysis. Is the diagram really structured? The entry and exit points are undecipherable; the hierarchical modeling is unthinkable. Figure 1.1.2C is a familiar sample of what has come to be known as "structured flowcharting." It is "structured" to the extent that each operation seems to have a single entry. But the single-entry device is controversial. A defect of this conventional "structure" flowcharting is that is may be misleading as to the *actual* entry paths. It has a tendency to show both single entries at the same point. We know this is not true in practice. A (conditional) BRANCH may not return to its point of "call" at the end of execution. In addition, its hierarchical structuring is questionable. But its strongest weakness is that it violates the laws of closure—a sine qua non property of algebraic systems. Is it really structured thoughtfully? How much time and effort will be required to decipher the "flow paths"? There is no doubt that there is fun in these diagrams, but can we always afford the extravagant time to produce them?

Once again, observe the influence of language on our thoughts. The FOR construct tends to lead us to this diagram type. Suppose we reconstruct our thoughts to enclose the ITERATION structure as a rectangular box; then, we can see the decisions of the nested-IFs as nodes on a binary tree. We simply draw them with their attendant hierarchy structure.

In an effort here to make this charting more logically controllable, easier to chart and convenient to use, we shall base our charting on a well-known structure—the binary tree. Figures 1.1.2D and 1.1.2E introduce the IF-tree diagram as a logical substitute for representing nested-IFs. It is important to realize that the nested-IF structure is a binary tree with obvious hierarchical attributes. A binary tree is a tree whose every node has two branches or is atomic (i.e., has no branches). Moreover, this device further enables better representation of the generalized SELECT structure in all of its total dimensions. However, the generalized select statement representation ceases to be a single binary tree; instead, any branch may become a binary subtree.

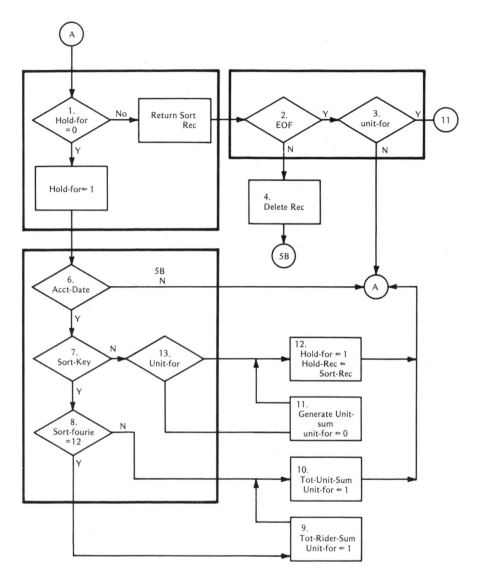

Figure 1.1.2A Traditional flowchart. The "zone" rectangles are drawn for enclosing "major" functions (as would be seen in a structured status).

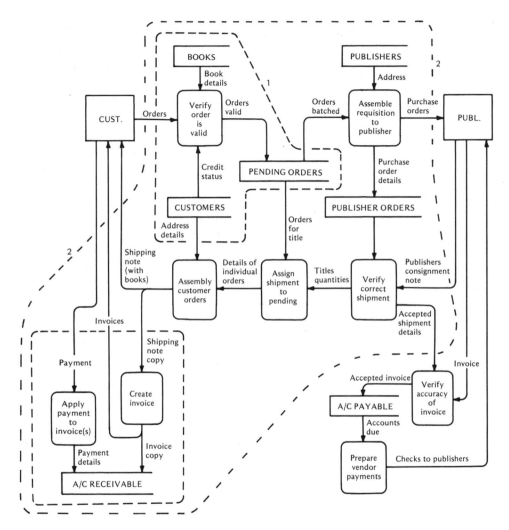

Figure 1.1.2B Conventional structured systems analysis diagram: a structured spaghetti?

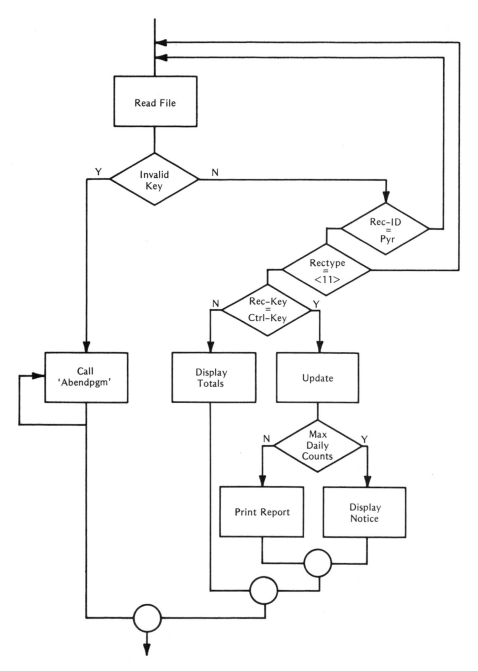

Figure 1.1.2C Conventional structured flowchart. Lacks structural hierarchy: On one side, several sublevels are "coalesced" into
equivalence with CALL-ABENDPGM. Repeated reading
of file violates hierarchy, giving a bottom-up substructure.

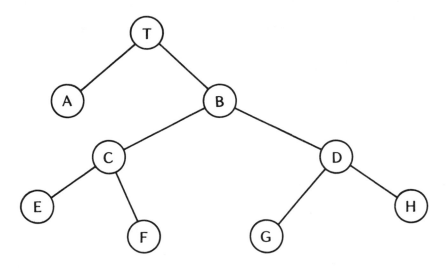

Figure 1.1.2D IF-tree diagram: actually, a binary tree.

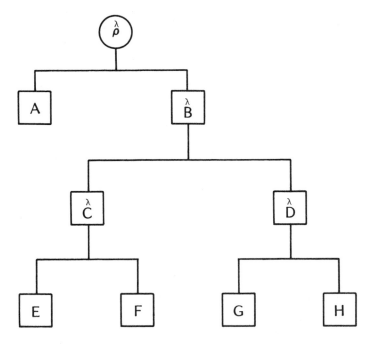

Figure 1.1.2E Annotated IF-tree diagram. The symbol, λ indicates a predicate test; boxes without λ are process functions.

The statement (shown in Figure 1.1.2F)

```
IF cond-A
   IF cond-B
      IF cond-C
         DO HIERARCHY-LEVEL-C
      ELSE
         DO HIERARCHY-LEVEL-D
   ELSE
      DO HIERARCHY-LEVEL-B
ELSE
   DO HIERARCHY-LEVEL-A.
```

translates to the diagram of Figure 1.1.2G.

The binary tree introduced here is the next fundamental step toward a comprehension of dimension in the theory of hierarchy.

In summary, traditional flowcharting techniques and pseudo-structured flowcharting can be better replaced with (easy descent) tree diagrams. The following best summarizes the frailties of flowcharting:

Program flowcharting violates the property of independence of functions; buries dimensionality; destroys rectifiability; obfuscates continuity and complicates control.

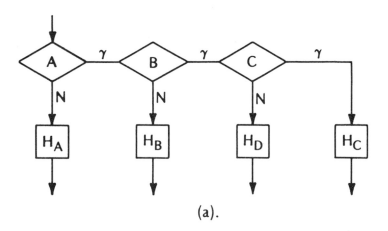

(a).

Figure 1.1.2F Nested-IF structure. Traditional unstructured chart.

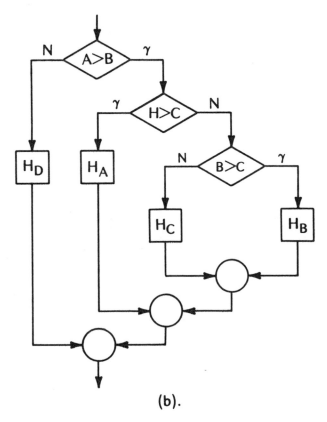

(b).

Figure 1.1.2F Conventional structured chart: the hierarchical structure seems to have been lost; and all the exits are routed to the same final point—a violation of the single-entry/single-exit axiom.

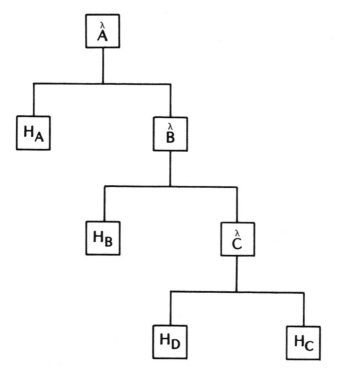

Figure 1.1.2G Nested-IF structure—equivalent of Figure 1.1.2F. The symbol λ indicates a selection and the boxes without it, actions.

The developments given in Part Two, which use the attributes of the tree diagram, will strengthen these arguments. With flowcharting methodology, the metric properties of software engineering are inconceivable and unobtainable. This explains the present chaos in programmer productivity studies. Considering the great value of tree diagrams, and the desire to postulate a scientific methodology that will be usable by all professionals (trainee and adept alike), it is critical to realize that conventional program flowcharting is inconsistent with structured programming.

1.1.3 THE SINS OF DOCUMENTATION

The sins of these spaghetti systems are conspicuously manifest in their documentation. Like a story without a head or tail, documentation for these systems have become burdens to programmers and analysts. It is possible these frustrations have forced some professionals not even to document some systems. Many programming centers have no documentation of some currently operating systems, and the documentation that does exist suffers from omission, redundancy or superfluity, or a combination of these. In other words, the documentation lacked purpose, exactitude or clarity, and thus defeated the purpose of documentation.

During maintenance or enhancement, these frustrations increase. Perhaps, it is this experience that generated the familiar bias the programming profession has against maintenance—a bias that has been exaggerated, perhaps because of the lack of reliable measures to assist supervisors in estimating the cost and duration of a maintenance project. The intensity of this bias has also disaffected new employees, who have come to share the same bias against maintenance. Its roots are grounded in the bitter experience gained from unstructured systems.

However, if practiced with structured techniques, maintenance can be very rewarding. In many departments, most of what the programmers do is work on existing systems—maintenance. By deciphering and improving what others have done, programmers enrich their own knowledge, as well as learn new structures and techniques. They become researchers. In addition, maintenance projects involve less defining of new files and setting up of a new job control language or work flow language, and actually takes much less time, giving the programmer a feeling of accomplishment.

Despite the proliferation of literature about structured programming, very little exists about structured documentation. A generalized approach in this direction (since it advocates the use of a project team and a review committee) appears in Teichroew [47].

Ordinarily, it is safe to assume that a structured system gives rise to structured documentation. However there is no absolute claim that structured documentation (i.e., documentation of a structured system) is always better than nonstructured documentation. Rather, the former has the capability of better aesthetics through structure-controlled detail and optimality. Structured documentation can also anticipate future modifications rather than merely satisfying current needs. Problems of cross-references, data dictionary, abstracts, indexes, memos, etc., are better managed in structured documentation.

Another advantage of structured documentation is minimization of the problem of understanding programmers' codes. The anarchy caused by undisciplined coding is inherent in the spaghetti system techniques. More recently, thanks to the revolution in data base management systems, computing centers are introducing standard naming conventions. The data dictionary (fig 1.1.3A) with its structured naming conventions, has become a significant technical addition to documentation.

Finally, it is common knowledge that the role of management in a systems environment (Figure 1.1.3B) can enhance or handicap (structured) documentation. Where management policies are explicitly defined, documentation will be at least close to being adequate. On the contrary, where management policy is ambiguous or even ill-communicated, documentation can become a patchwork of confusion which will manifest itself during system maintenance. The solution to this is the use of chief programmer team and walk-throughs.

Ser. No.	Name	Type	Size	Mode	X–ref	Miscellaneous

Figure 1.1.3A Data Dictionary—an index of data which enhances data independence and uniformity of control

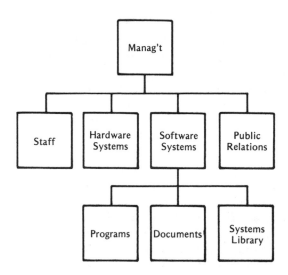

Figure 1.1.3B   Management in software engineering—familiarity with methods and metrics can lead to better productivity

Since documentation is considered a secondary but simultaneous pursuit of software engineering,* its theory is not pursued in the rest of this book. It is expected that the trends and impacts of structured programming will be reflected in structured documentation.

1.2 THOUGHTS, SPECULATIONS, PAPERS AND LETTERS

Given this state of confusion and skyrocketing production costs, a number of computer scientists began to seek ways of refining and improving this situation. A potential handicap toward achieving the goal of this research turned out to be the wide gulf between academia and industry. Research papers published in journals are read only by a few; and usually, these papers are either not made available to industries or industrial technocrats do not make the necessary effort to know what research scientists are thinking or publishing. In many programming environments, the only

---

*The author considers software engineering a triad of structured programming, structured documentation, and structured management.

reading materials provided are programming manuals and operating journals. In other words, industrial application traditionally lags behind research postulations.

Most traditional programmers give very scanty regard to mathematical reasoning (Figure 1.1A) for designing systems. The general public usually regards mathematics as only for computing numbers, and not as something dealt with in the daily activities of life. The fundamental usefulness of mathematics is that it offers us a chance to prove our constructs correct or incorrect, even in simple activities. We shall see this in chapter 5. For now, let's look at some influential thoughts on this.

First is the proof by C. Böhm and G. Jacopini [5] that any program structure can be expressed as one of or a combination of the following building blocks:

1. *Sequence*—a set of instructions executed in the order written: first, second, third, etc.

2. *Selection*—IF-THEN-ELSE, a structure that depending on condition-clause enables the choice of execution of sequence paths.

3. *Iteration*—DOWHILE or DOUNTIL, a structure that enables the repetition (looping) of some sequence, while a certain condition is yet active.

The first alert to software professionals about spaghetti systems came from professor Peter Naur in 1963. However, according to Knuth [31], who traced the history of the GOTO controversy, the first programmer who "systematically began to avoid" GOTO statements was D.V. Shore in 1960. Then in 1964, George Forsythe was purging GOTOs from his algorithms, though not completely. In 1965, Edsgar Dijkstra published his communications about Algol 60 in which the GOTO statement was abolished. Other contributors toward a disciplined use of GOTOs include Peter Landin [32], Heinz Zemanck, Christopher Strachey, and C.A.R. Hoare.

Several other researchers were initiated around this same time, especially in the area of logic and computability [6] of programming languages. Some of the researchers developed programming languages that did not use the GOTO statement at all, such languages as LISPX, MOL–32, ISWIM [52, p. 145].

Top among these speculations was the need to develop a "scientific" methodology for proving the correctness of programs. Despite several efforts in this direction, no clear solution has yet emerged. The illusion still lingers, for according to Dijkstra, a proof of something is never an end by itself. Suppose a postulate has been proved correct; then its proof must be verified, thus leading to another proof of correctness which must itself require another proof of correctness. This circle is an infinite process [18].

Associated with this dilemma of proving the correctness of a program is the problem of an effective theory of program testing. Several research institutions, computer manufacturers and universities have invested considerable efforts toward a solution. Much remains yet to be done, but there does exist interesting research results and literature [38, 25].

A major paper by Joel Aaron of IBM, *Super Programmer Project Experiment* [1], offers the concept of chief programmer team, CPT (changed from superprogrammer). Papers by Baker and Mills [3] accentuate the success of IBM in structured methodologies [2]. Though CPT is purely a managerial methodology, many agree that its impact was immediately instrumental in pushing the structured revolution to public acceptance.

Some other remarkable speculators and contributors to this creative revolution deserve mention here. The works of Yourdon—books [52, 53], letters, lectures and seminars, including video tapes—cannot be omitted. Many professionals are also familiar with Wirth [49, 50] (inventor of Pascal), Knuth, and Chapin and his popular *Chapin Charts* [8, 9].

In summary, one can conclude that these and other scholars and researchers have contributed to building structured programming on a triad of concepts:

1. GOTO-free programming
2. Efficient and economical constructs with the fundamental structures of sequence, selection and repetition
3. Hierarchical structuring for well-controlled refinement of complexes up to minimality.

1.3 DIJKSTRA'S FAMOUS LETTER

The impetus for popularizing the structured revolution is credited to Dijkstra of the Netherlands. His famous letter in the *Communications of ACM* [17] in 1968, entitled "GOTO Statement Considered Harmful," persuaded many computer scientists but also puzzled some. Unfortunately, this letter received limited circulation, mainly among academics.

However, it generated several articles [14, 31, 33, 35] from readers and thus brought the message home to many. Dijkstra admits his thinking was influenced by such elites as Hoare, Wirth [49, 50], Strachey and Landin [32, 45].

In practice, it must be recognized that the GOTO consciousness was originated by Naur. The following argument best describes the crisis [56, p. 36]. If an algorithm, A_1, uses a GOTO construct to transfer control to another algorithm, A_2; and if A_2 repeats the pattern to A_3 and the chain of transfers continues to some A_N, then these algorithms clearly become separated in space. (The printed listing appears on separate sheets.) This imposes a hardship on tracing the behavior of the program's flow. Worse still, during maintenance this already complicated debugging further hinders modifications. In a nutshell, the GOTO introduces a loss of control which causes debugging to stagnate; debugging, in turn, generates more and more maintenance, leading to a complex patchwork.

The primary message of Dijkstra's letter was that the quality of programmers was inversely proportional to the number of GOTO statements in their programs, that is, as the number of GOTOs increased, the effective quality of the program and therefore the productivity of the programmer decreased. Undisciplined use of the GOTO statement may cause the corresponding reference point for the GOTO statement to become lost, thus making the program flowpath difficult to trace.

Dijkstra's first message in this letter accents the importance of the dynamic processes taking place under the control of the program away from the programmer's presence, i.e., at execution time. He contrasts this with the static process which comprises the relationships between data and operands.

Dijkstra's second message points out the limited capability of humans to understand rapidly changing processes. His argument is that humans are traditionally disposed to master static interactions, but are less attuned to comprehend dynamic processes [19]. In the letter we find the body of Dijkstra's contribution to the structured revolution:

1. Disciplined use of the GOTO statement.

2. Use of the hierarchical concept for the design of program logic under the control of sequence, alternation, and iteration clauses.

3. Use of the concept of *elegance* to enhance the proofs of a program's correctness.

4. Arguments against von Neuman type machine code—the ability of storages to modify themselves.

1.4 ARGUMENTS AGAINST
STRUCTURED PROGRAMMING

———————◆———————

Although the opposition to structured programming continues to wither, there are still scattered pockets of resistance. Therefore, let's discuss some of the causes and sources of opposition to the structured revolution.

Traditionally, a major impediment to change or progress is contentment with habit or satisfaction with orthodoxy. Most people do not readily discard well-known habits in preference for new ideas. Those who have become addicted to flowcharting are putting up the stiffest opposition; this is understandable since the primary element of flowcharting (GOTO) is under attack. Combined with this contentment with habit is the general human reluctance to retrain. Naturally, the structured programming revolution will call for more classes, seminars and lectures and this could perhaps influence promotions.

Associated with this is the gulf between academics and professionals. Early speculations, discussion and experimentation about structured programming were conducted by academic people and were published in scholarly journals which were outside the reach of the average software professional. Most examples used scientific applications and programming languages. Many of the books written for the commercial or business applications aimed at simplifying rather than exploring the problems of programming; they were targeted for immediate application rather than comprehension. In addition, the academic world very often naturally indulges in experimentation; the business environment wants instant results.

It may be argued that part of the repulsion to academic innovations in concepts is the manner in which academic proponents clothe old ideas. Such concepts as functions, recursion and modelling are really elementary, but their lecture presentations sometimes repel the commercial programmer. The lesser sophistication (mathematically) of commercial programs has deluded some "well educated" individuals who feel that this level of computer programming is for high school graduates. The commercial environments were pressured into producing immediate results. There was no time for study of methods, and very few people had had any classes in programming. In professional environments, the concept of chief programmer team raised eyebrows—how many programmers are really super, in how many installations and in what applications?

Another argument against structured programming is that it produces *more* coding. Although the influence of the programming language cannot be ignored, the author's experience showed unequivocally that structured techniques, if used well, actually reduce the amount of coding. In the particular case, the author eliminated the THRU option of the PERFORM statement (of Cobol language). The result was that the procedure names listing was halved, and the source listing was 25% smaller, in addition to reducing the core size. It seems then that the degree of structuredness (or optimality) greatly influences the size of coding. What one programmer codes in 400 lines, another may code in 250 or even less lines. The belief that structured coding can lead to duplication of codes is also solved by the degree of optimality. A well-structured program actually uses *less* lines of code than its contrary.

One common complaint against structured programming is that it increases inefficiency. The two victims of this are memory size—because of duplication of codes—and CPU time—because of several calls to subroutines. This may be so because of excessive use of CALL mechanisms by programmers.

More than a few CALLS in the same module are symptomatic of poor module logic. They frequently lead to wasteful input/output processings (worsened by the number and sizes of calling parameters or chains of calling modules) and thus to excessive execution time. Such modules simply perform very inefficiently. Only by optimizing the design can such CALLS be minimized.

However, it cannot be denied that certain matters of efficiency and elegance in performance may sometimes necessitate duplicate coding. In these cases, efficiency due to savings in execution time (for virtual memory, especially) override economy due to savings in code. A typical example comes from Fortran IV where we have to initialize several arrays:

```
      DO   88    J    =  IN, IT
             ANN (J)   =  0.0
             PET (J)   =  0.0
             CAT (J)   =  0.0
             KEN (J)   =  0
 88   CONTINUE
```

This coding has the potential of causing a page fault if the matrices are large and exist in noncontiguous storage locations. A better technique is to code separate loops for each array [46].

There is also a very genuine complaint that structured programming uses too much development time. This is because structured programming strives for efficiency and elegance—what is worth doing takes time. We may argue that the "long" time taken for development (design, code, compile, test) is for educational orientation to this new philosophy. A few years hence when programmers will have absorbed more of the principles of this philosophy, they will perform better and faster than they are now. In practice, the long development time is offset by shorter testing times and minimal debugging frustrations. Any methodology that leads to less debugging is obviously a thousandfold better than one that consciously or unconsciously buries bugs into systems. Paralysis in debugging can overwhelm the cost of prolonged but controlled design cost. Worse still, it can encourage high turnover.

Despite these and other similar objections, and realizing that the structured revolution swiftly and uninterruptedly blazed the trail of information management, we must conclude that the consequent confusion and controversy have suddenly raised the structured programming principles from a tentative status to one of acceptance and enforced application [56, p. 11].

Finally, we close these arguments with the somewhat "paranoid belief" of a seasoned professional and writer. According to him, structured programming, regarded as a "fancy dressing" by some and a "mild mannered improvement over traditional practices" by others, is a "fascinating plot" as well as "a political manouvre" which, like other efforts to invent programming languages to replace Cobol and Fortran, has proved "a failure at what it set out to do." [22, p. 28]

EXERCISES

1.1 (a) The most prominent violation by spaghetti systems is against the concept of *sets*, in particular, equivalence relations on sets. Show that (traditional) flowcharting emphasizes logical rather than program flow structure.

(b) Give reasons to document the uses and advantages of documentation. Recently, documentation is receiving popular attention in the industry and through professional seminars. Show the influence of documentation on maintenance, a corporation's operating functions, and programmer productivity.

(c) Identify the differences between traditional and structured methodologies.

1.2   Identify any useful *idea* for improving programming techniques not covered by current speculations and letters.

1.3   Dijkstra's analogy of "pearls on a necklace" as well as his concept of hierarchy were very influential in his invention of the terminology "structured programming." Show that hierarchical thinking is a powerful practical tool for the management of complexity.

1.4   Find more criticisms of structured programming. Justify each in regard to the total set of criticisms.

Chapter 2

MANAGEMENT
of STRUCTURED
PROGRAMMING

Management, in many of its facets, is about the oldest institution in every industrial concern. However, the people are still crying for more management, and in this age of structured thinking have called for structured management. By this they mean a well-defined purpose for the accomplishment of a visible attainable goal along a consciously directed path. This is quality management.

2.1 WHAT IS STRUCTURED PROGRAMMING?

Most great revolutions descend with varying climates of ideologies. With them emerge prophets, gurus, salesmen, experts, exploiters, etc. To the extent that the revolution succeeds depends on the strengths of these armies. Each expounds its own way and, more than that, attempts to deny others their way.

Over a decade since its emergence on the programming industry, there is no concensus as to what structured programming is. Instead, there are diverse concepts and practices all going under the name of structured programming. This chapter will separate managerial techniques from methodical ones and discuss them all.

An experiment at IBM combined effectively two rational methodologies—management (of programming) and technical (structured) application. The combination of professionals drawn from each of these two fields had great rewards for increased productivity. The group was called the Chief Programmer Team. Like a well-trained and well-practiced team, their task was to produce the most from the minimal resources available.

However, it must be borne in mind that the chief programmer team by itself is not exclusive to structured programming. It is just a tool. To an extreme, structured programming as a methodology can be accomplished effectively without resorting to the organization of a chief programmer team. The concept and practice of the chief programmer team is purely a managerial approach. Its great merit is effective control of the project. In fact, Mills proposes an analogy between a chief programmer team and a surgical team [2] (consisting of specialist doctors, nurses, anesthesiologists and laboratory technicians).

Another analogy to the chief programmer team is found in academic research. A group of students taking the same subject may be organized to do research on a certain topic. One among them is selected to lead the group; of course, he or she must be sufficiently knowledgeable about the topic. Then some others are selected to do the library bibliography and all associated clerical work, while the rest are selected to do the actual collection, redefinition and proving of theorems or propositions. This group clearly behaves like a classical chief programmer team.

2.2 THE STRUCTURED REVOLUTION

———————◆———————

Every theory has its corresponding practical application. Historically, application most often is the sequel to theory, for the ultimate acceptance of a theory is its demonstrated value in application. But in actuality, application precedes theory. The person who theorizes must have amassed some questions, answers, puzzles and skepticisms that consciously or unconsciously triggered the moment of theorizing. Those who pioneered the path to structured programming did so out of their own accumulated experiences and conjectures.

Structured programming would have remained behind academic curtains and industrial cobwebs, even until today, but for the courage of some research scientists who decided to put to *test* what had long been committed to thoughts and speculations. The announcement of IBM's successful experimentation with structured programming (The New York Times Project and the Definitive Orbit Determination Project) attracted considerable attention from users and researchers. We can say that this gave the green light for the adoption by industries of structured methodology and the contributions of Dijkstra and several others.

The foundation for the technical architecture of structured programming was laid in a paper by Böhm and Jacopini published in 1966, entitled

Flow Diagrams, Turing Machines and Languages with only Two Formation Rules [5]. Using the methods of mathematical reasoning, they showed that any program can be written as a sequence of only three basic building blocks:

1. Sequence—program statements placed in their order of execution.
2. IF-THEN-ELSE—allows groups of statements to be included in the THEN and ELSE paths; these groups of statements may be any of (1) or (3).
3. Loop control mechanism—allows iteration or repeated execution of an especially structured sequence of statements. The control mechanisms are DOWHILE or DOUNTIL.

Actually, a program is a sequence of algorithms. One significant attribute of the Böhm-Jacopini theorem is its independence of programming languages. Thus, structured programming transcends programming languages. (Except for a few examples, we shall maintain this philosophy throughout this book.)

However, the Böhm-Jacopini paper does not attempt to show that these basic constructs are the best or the only forms for constructing programs. As Denning has said, an open-minded programmer may not reject any addition to this basic set, provided such an addition is simple, understandable and satisfies a well-defined proof [56, p. 39].

Another powerful concept for structured programming was introduced by Mills of IBM. He extended the above results by the additional requirements of single-entry/single-exit [36]. The importance of this innocent-looking concept cannot be overestimated. It is a disciplined link for the flowpaths of two or more of the basic blocks. Consequently, we shall show later that this feature belongs to the theory of hierarchical structure, rather than to a fundamental structure (of the Böhm-Jacopini type).

In summary, one can argue that the "chief programmer team" for propagating the theory and practice of structured programming comprised the following individuals:

Chief Programmer—H.D. Mills

Backup Programmer—E.W. Dijkstra

Programmers—Böhm and Jacopini

Librarian—F.T. Baker

2.3 CHIEF PROGRAMMER TEAM

———————◆———————

At a 1969 NATO conference on software engineering techniques [1, 7],
Baker and Mills of IBM described a successful experiment undertaken
by IBM. It was called the super programmer project, used on Skylab's
definitive orbit determination system. This system required the use of
sufficient mathematics to develop the necessary functional equations.
The purpose was to use a small group of people constituting the chief
programmer team to do the work that would have required a larger team
under the traditional methods of program development.

Historically, this is the first application of this managerial technique
to production programming. It was the best thing that ever happened to
programming—complimenting operational expertise with management
know-how. Its impact, particularly when combined with walk-throughs,
is destined to eliminate many managerial handicaps. Traditionally, programmers had experienced very isolated contacts, even when working
on different components of the same system. It has been "code your
own thing." The consequences of this loose coherence in management
have untold failures, particularly at testing.

The primary team members (see Figure 2.3A) in a CPT organization
are:

Chief programmer

Backup programmer

Programmers

Librarian or secretary

Most authors don't include programmers, but it is done here for com-
pleteness. One other aspect mentioned by popular writers is the size of
the CPT. If the number of lines of codes exceeds 100,000, they recom-
mend increasing the size of the team. This is confusing since we have
to know the number of lines of codes before deciding on the availability
of people. The size of the team should depend on the magnitude and
complexity of the project, as well as the size of the corporation. In other
words, the size of the team grows as more is learned about the complexity
of the project, typically after the analysis phase is concluded. Finally,
another confusion associated with CPT is the technique of top-down.

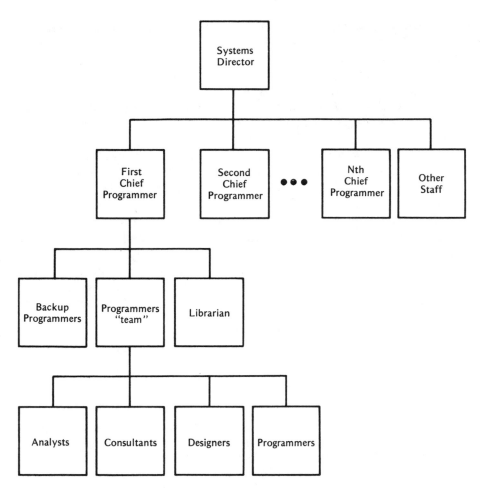

Figure 2.3A   Chief Programmer Team

Top-down is an operational know-how required of the programmers in the CPT. Its relationship to CPT is the mere decision to adopt it. It is not a managerial tool.

The *chief programmer* is an experienced programmer with some management know-how as well as thorough familiarity with the project. He or she is the technical manager, overseeing development, definition of design elements, coding, testing and integrating of all the modules or interfaces.

The *backup programmer* is nearly as experienced as the chief programmer. In the absence of the latter, he or she assumes full responsibility for the project and should fill in all necessary details to supplement the task of the chief programmer.

Under *programmers* are included designers, analysts, programmer-analysts, programmers, and all professionals and specialists. The number of them depends on the size and criticality of the project.

The *librarian* or secretary assumes the clerical responsibilities of the task—maintenance (including updating) of files, codes, library modules, program listings, data dictionary, documentation, etc. The librarian need not have a strong technical background.

2.4 STRUCTURED WALK-THROUGHS

If the chief programmer team application enforces control and discipline on programming, walk-throughs inspire confidence and creativity in programming productivity. If used effectively, production costs will greatly decrease and many of the irritations of programming will be alleviated. But, on the other side, walk-throughs can *generate* irritations. Many programmers are not trained in the scientific way of sharing and giving arguments required in walk-throughs without becoming uncomfortable. This is why the traditional "code your own thing" has persisted. Walk-throughs really bring programming into the fold of scientific methodology.

Walk-throughs were developed by IBM as part of its CPT concept and practice and are technically a pre-test debugging device. It is essentially a generalized desk-checking technique, utilizing the experience, observation and expertise of all the team members. In a walk-through, the analysis, design, code or test versions of a program (project) are reviewed under a set of rules drawn up by the organizer—usually, the chief programmer or a peer. Copies of the listings (of design, code, test) are distributed some days before the scheduled meeting to allow participants to pre-read and formulate questions, suggestions and criticisms. These practices do more than break down the myth that a program is the exclusive personal belonging of the programmer—they assure all concerned that a walk-through is an "open heart surgery" of the program, an environment for rational give and take for the good of the group. Its sole objective is to reinforce confidence and cooperation.

In a walk-through session, one person is appointed to act as the secretary, taking notes of comments, suggestions and changes as well as recording other useful information about the session [22, C.2.1].

This report is annotated and distributed to all participants hours after the session. This serves as notice of corrections or amendments to be made to the program. It is the special responsibility of the programmer to include these corrections. There may be another walk-through session after these corrections, depending on the gravity of the errors found. Some specialists or even observers may be invited to be present. It is their privilege to resolve any disputes that may arise between the chief programmer and other participants. In practice, it is very useful to have junior programmers attend walk-throughs as nonparticipating observers for the sake of exposure. An important member of the staff who must be *excluded*, to the satisfaction of participants and programmers in particular, is the programming manager or systems director. Programmers unused to the scientific argumentative way may fear the manager is there to judge (for promotion). This fear justifies his or her exclusion.

In conclusion, even if the chief programmer team approach is not utilized, a walk-through is very highly recommended, irrespective of the size of the programming environment.

2.5 MANAGERIAL TOOLS

Though the theory of management is completely outside the purview of this book, there are other managerial tools and concepts that may be incorporated into the management of structured programming. Some are exercised beyond the activities of the chief programmer, and directly by the center's top management. For instance, the systems director handles personnel matters for the chief programmer team, since the chief programmer is *primarily* concerned with the technical direction of the project.

Concepts and practices which ensure efficient and effective management direction are: project review and control, checkpoint, monitoring, coordinating, budget control, and use of specialists when necessary. However, these tools must not be used in such a way as to jeopardize the central goal. Some of them are really not necessary for certain sized projects.

Other aids for managing structured programming include some software packages—structuring engines; these are included under managerial techniques, since their use falls outside the scope of CPT and the Böhm-Jacopini and Dijkstra-Mills bases principles. Although the author has not used any of these tools, his experience with manual structured designs arouses skepticism about their effective usefulness. The author knows of no automatic device that proves mathematical theorems, whether familiar or yet undiscovered. By the same argument—and repeating that the basic building blocks of the Böhm-Jacopini theorem were founded around mathematical reasoning—the author knows no rational, feasible, provable, satisfactorily automatable technique for structured programming. Some of these devices may actually transform GOTO statements into other structures, but this is built on a false assumption that structured programming is just the elimination of GOTO statements or perhaps indented coding. Chapter 5 will demonstrate that well-structured programs cannot be obtained by the mere formalistic elimination of GOTO statements [47]. There is a definite human intellectual ingenuity to be employed in designing programs from original thought processes, rather than transcribing them from other program forms.

Finally, management reviews may also include functional specification, design, code, documentation, test strategy, training materials, procedures and standards, operations, maintenance, etc. [22, Part F].

EXERCISES

2.1 Extend the list of notions about the structured revolution and criticize each practice. Discuss the revolution's usefulness and place in structured programming.

2.2 (a) State the importance of management in systems development (include maintenance).

(b) Identify any factors you consider consistent with effective management of EDP productivity.

2.3 It was stated that the CPT is sufficient but not necessary for structured programming. Criticize this argument. What responsibilities do you consider adequate for an efficient CPT?

2.4   What responsibilities would a walk-through committee undertake? Give reasons to demonstrate the advantages of a walk-through in software production.

2.5   Identify and describe other management techniques or tools for improving programmer productivity.

Chapter 3

AESTHETICS of STRUCTURED PROGRAMMING

In the beginning the structured revolution itself lacked structure. Most of the pioneering ideas were tossed off as conjectures, only to be later enunciated and formalized. There were no axioms, no well-defined principles and no theorems. The formalization of these and other notions are just beginning to take root. This chaos allowed for a proliferation of notations, concepts and practices. Even many "structured" programs turned out to be nests of bugs. To improve readability, ignoring for now correctness of proofs, much attention was focused on manufacturing good codes; first there was pseudo-code, now called semi-code or structured english [8]. The philosophy of structured coding is still growing, beyond coding program instructions to writing specifications. These successes are not sufficient; we must also comprehend the principles of structured programming.

3.1 SIX AXIOMS FOR STRUCTURED PROGRAMMING

———————◆———————

Structured programming begins with the analysis and design phases of the program, for without a structured design, it is very unlikely a structured program will be created. There is also structured specification, structured documentation and structured maintenance. Consequently, the concept of structured programming permeates the system's life cycle. Rather than trying here to answer the question, "What is structured programming?" we might benefit more by tackling the question, "When is a program (well) structured?" By working with a proclaimed "structured program," we shall examine its attributes to determine whether or not it is structured. We shall assume a set of axioms and determine how the given program satisfies these axioms. (Some of these axioms will be refined and presented in chapter 5 as (accepted) principles.

Let's proceed as follows:

1. We expect the program has been assembled from discrete pieces of instructions; several pieces may be arbitrarily combined to obtain what will be called a *net*, a set of homogeneous instructions.

   Axiom 1: Program consists of nets of instructions.

2. We shall make a sensitive and dramatic concurrence: Global program flowcharting has been outlawed owing to ulcers it inflicted on several programmers; program flowcharting encourages GOTO constructs, in addition to violating hierarchical structure. The reason for outlawing it is that we want the program to read sequentially for immediate understanding. By "read sequentially" we mean that there are no sudden hurdles to jump.

   Axiom 2: Every net (program structure) is GOTO-free. Some programming languages may require a weaker form of this clause.

3. A member of a net may make reference to a point outside its net, but in doing so it should not get outside its net arbitrarily. Every reference to a net *must* reference a common point of entry and a common point of exit; these points are common to those two nets. In other words, each net behaves like an electrical circuit. It receives electric current at one and only one point and conducts it away through exactly one other point.

   Axiom 3: Every net has exactly one point of entry and one point of exit.

4. We shall require that the whole program be built of *sequences of nets*, in such a way that starting at the top net, we begin with the first sequence: read all the statements of this net from the top through to the last "branch"; next, from the top, we do the same with the second net, and the third net, and so on, until we finally terminate at the last net of this last sequence of nets. In a sense, the sequences of NETS are hanging vertically downwards, like an inverted tree.

   Axiom 4: Nets communicate with each other in a super- and subordinate fashion.

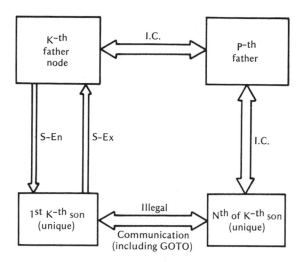

Figure 3.1A Characteristic interactions of the axioms for structured programming. S-En = single-entry; S-ex = single-exit; I.C. = illegal communication

5. A net may require a fixed number of traversals greater than one. But this does not alter its contents of sequence statements. Such a decision to traverse (repeatedly) may be controlled in a net directly above it but on the same chain of sequence of nets.

   Axiom 5: This super-to-subordinate communication is one of a father-to-son relationship.

6. Finally, we require the contents of any two arbitrary nets, chosen from any arbitrary levels or coordinates, to be unequal.

   Axiom 6: Every net belongs uniquely to a point in the program's hierarchy.

The picture we have just illustrated demonstrates how to organize the structures that comprise our program. It therefore has inherent in it both how to design and how to analyze for constructing our program. It enables us to program consciously using at each step a minimal set of controllable program structures. This is the kernel of structured programming.

What these structures are will be introduced in chapter 5. The rest of this chapter will discuss those additional attributes that are sometimes erroneously judged as structured programming.

3.2 STRUCTURED CODING

————————◆————————

The general literature in current circulation uses the terminology "structured coding." However, structured coding is but one aspect of the application of structured programming. The reason is again inherent in the confusion about what constitutes structured programming. Some people argue that structured coding is the coding of structured programs, but the effort here has been to separate substance from shadow. The basic principles of structured programming must not be confused with their usage. Application is not principle. Here structured coding will receive a more restricted meaning and will include such topics as:

1. Indented coding

2. Delimiter coding and blank line insertions

3. One-line per condition IF-THEN-ELSE coding, etc.

4. Data naming conventions

The inspiration to indented coding was the need to improve program readability. Program readability contributes to better proofs of program correctness, which, in turn, contributes to improved program productivity. One of the basic skills in which a programmer should be trained is program reading. The ability to read another's program is proof of the capability to change, modify or improve programs. Just as good editors, journalists and engineers learn from criticizing other works, so must a good programmer. Since nearly every motivation for the structured revolution revolves around proving program correctness, it is clear that almost every resulting effort in structured programming is regarded as an aspect or even a variation of structured programming.

Programs in some of the traditional programming languages allowed little or no blanks in between two words. For instance, in Fortran, one could code:

```
IF(KC.NE.JD)GO TO 25
WHILE(N.NE.0)DO
SUM = SUM + 1
```

In Cobol blanks are, by definition, imperative between any two words. But structure-minded programmers find it more readable to use two or more blanks instead of one. Another problem is that of coding two sequences of IF-THEN-ELSE on the same line or joining the ELSE portion to the same line that contains its IF construct. The author has seen several programs (Fortran and Cobol) without even a single blank line. Two such programs each had over 4500 lines of code from IDENTIFICATION DIVISION to the last statement of the PROCEDURE DIVISION. Constructs like

```
IF condition-A THEN condition-B ELSE condition-C
```

in structured coding would be

```
IF    condition-A
      THEN   condition-B
ELSE
      condition-C.
```

Such a device allows easy recognition of flowpaths as well as distinction between predicate test and conditional action. This is the desired easy readability condition. The reader may also notice the extra blank provided between two words, called delimiter coding.

Structured nested IF-THEN-ELSE constructs exhibit a mental pushdown structure that enables one to maintain a proper coordinate position. They are actually a binary tree. This is the meaning of the concept of dimensions in design and programming of systems. This concept enables us to perceive the totality of a structure or object rather than seeing it in parts; we see all the parts at once. A tree diagram drawn on paper is two-dimensional; when we consider the notion of dimension, we come close to the actuality of a natural tree—branching out freely into N-dimensional space.

EXAMPLE 3.2A: Unstructured Nested IF-THEN-ELSE

```
IF  AG  >  BG  IF  AG  >  CT
                      MOVE    AG   TO   GREATEST
                ELSE  MOVE  CT   TO   GREATEST
    ELSE IF  BG  >  CT
          MOVE    BG   TO   GREATEST
       ELSE   MOVE    CT    TO   GREATEST
```

EXAMPLE 3.2B: Structured Nested IF-THEN-ELSE

```
IF   AG  >  BG
     IF  AG  >  CT
         MOVE  AG  TO  GREATEST
     ELSE
         MOVE  CT  TO  GREATEST
ELSE
IF  BG  >  CT
     MOVE  BG  TO  GREATEST
ELSE
     MOVE  CT  TO  GREATEST
```

For flowcharts of the unstructured and structured equivalents of these examples, see Figures 3.2A, 3.2B, and 3.2C. Notice the simplicity of flow and structure in the second example.

Finally, let's mention briefly the practice of data naming conventions. Some programming languages restrict data names: Fortran to six and Cobol to thirty characters. With the advent of data base management systems, individual computing centers are adopting different but convenient (fixed) data names. This allows for uniformity and maintainability as well as accelerated coding and debugging. A useful practice, for instance, is to associate the name of each datum with a major file that most uses it. A practice that allows portability is predefining common data structures and storing them in system libraries for copy availability. This promotes easy compilation and testing (provided such data structures are precompiled before storage in the library).

The use of qualified names (Cobol, in particular) has the obvious effects of excessive coding and debugging. The uniqueness of data names is simpler, straightforward and requires direct reference and less coding.

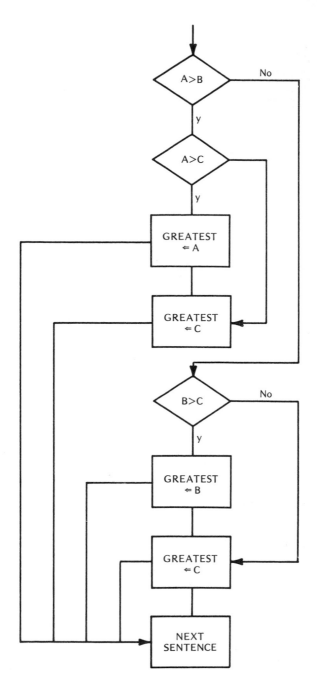

Figure 3.2A   Unstructured nested-IF

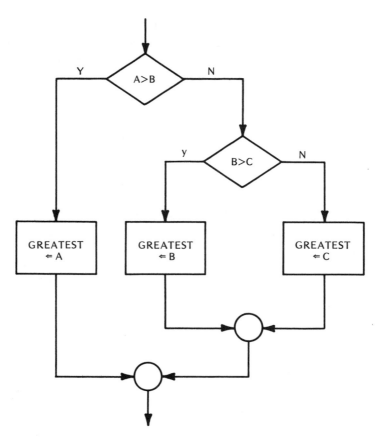

Figure 3.2B   Structured (conventional) nested-IF

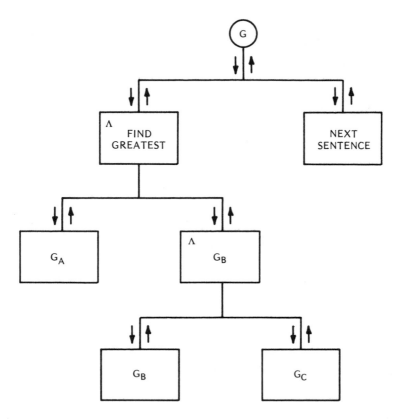

Figure 3.2C Structured (annotated) nested-IF. The symbol Λ indicates a selection of only *one* of the function-boxes immediately below. The boxes without the symbol Λ are "pure" sequence actions. The arrows indicate entry/exit for the corresponding function box.

3.3 STRUCTURED LABELLING

———————◆———————

There are structured programming advocates who are opposed to label-ling, notably PL/1 programmers. While this argument may have merit, the value of labelling can be seen in other respects. First, if data structures can be given names or labels, the same could be true of those function structures or nets which comprise the various sequences of statements that are the heart of structured programming. Second, these labels may serve as the logical entry points of the various function structures, some of which are subroutines. Third, a module (bearing an identifying name or label) may also be regarded as the highest label in a tree of labels—the module itself is the tree. A fourth value of labelling will be demonstrated below when the notion of dimension is introduced.

The structured labelling concept interferes with the basic structure of some programming languages. PL/1 or Algol has no place for it. Fortran, Cobol and BASIC have provisions for labelling structures and although programmers have been using labels, they are not generally familiar with structured labelling. Cobol programmers use something like structured labelling but very inadequately; BASIC and Fortran programmers are the worst offenders. However, because the whole philosophy of structured programming is independent of any programming language, this book will include a brief discussion of structured labelling.

Programming languages are still evolving and the structured revolution is having a dramatic impact on them. Such languages as Fortran (Fortran IV)—reputed as being unstructurable—has been undergoing structuring innovations; the same can be said about BASIC. The September 21, 1981 issue of *Computer World* announced Protran as a structured version of Fortran which was going to revolutionize the Fortran industry. With structured labelling, the tree of systems concepts will have as nodes the labels of the given system (module); this is one great advantage. These structure labels are analogous to the paragraphs of a prose written in human languages. In effect, the structure labels identify the larger structures (such as the SEQUENCE and ITERATION constructs) upon which hang the sequences of expressions. In addition, it is also a useful aid to readability. This is accomplished with the sequencing of nodes by levels in Examples 3.3A and 3.3B.

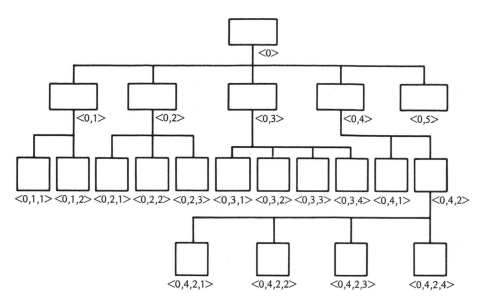

Figure 3.3A

EXAMPLE 3.3A: Structure label—Cobol (Figure 3.3A)

```
PROCEDURE  DIVISION USING  PARM-OK.
00-VER15MO7-CASH-ADVANCE.
    PERFORM   01-INIT-AREAS.
    PERFORM   02-INPUT-CARDS UNTIL  CARD-END.
    PERFORM   03-UPDATE-MSTR-FILE UNTIL  UPDT-END.
    PERFORM   04-CALL-REPORTS.
    PERFORM   05-CLOSE-FILES.
    STOP RUN.
01-INIT-AREAS.
    PERFORM   011-OPEN-FILES.
    PERFORM   012-ZERO-TABLE.
02-INPUT-CARDS.
    PERFORM   021-READ-CARD.
    PERFORM   022-XCHEK-CARD.
    PERFORM   023-STORE-TABLE.
```

```
03-UPDATE-MSTR-FILE.
    PERFORM   031-READ-MASTER UNTIL  CARD-REC.
    PERFORM   032-MATCH-CARD-TABLE.
    PERFORM   033-UPDT-REC-TYPE.
    PERFORM   034-RELEASE-MASTER.
04-CALL-REPORTS.
    PERFORM   041-SORT-MASTER.
    PERFORM   042-ISSUE-REPORTS UNTIL  LAST-REC.
05-CLOSE-FILES.
011-OPEN-FILES.
012-ZERO-TABLE.
021-READ-CARD.
022-XCHEK-CARD.
023-STORE-TABLE.
031-READ-MASTER.
032-MATCH-CARD-TBLE.
033-UPDT-REC-TYPE.
034-RELEASE-MAST-REC.
041-SORT-MASTER.
042-ISSUE-REPORTS.
    PERFORM   0421-RETURN-REC.
    IF RET-REC-TYPE = 'A'
        PERFORM   0422-CAL-DEPT-A
    ELSE
    IF RET-REC-TYPE = 'B'
        PERFORM   0423-CALL-DEPT-B
    ELSE
    IF RET-REC-TYPE = 'C'
        PERFORM   0424-CALL-DEPT-C.
0421-RETURN-REC.
0422-CALL-DEPT-A.
0423-CALL-DEPT-B.
0424-CALL-DEPT-C.
```

EXAMPLE 3.3B: Structure label—BASIC (Figure 3.3B)

```
00  REM PROGNAME = ESPBAS00
    OPEN 'DATADICT' REC 50 AS 01
    OPEN 'FORMDIRC' REC 80 AS 02
    G  = 2.5
    J% = 1
    K% = 1
```

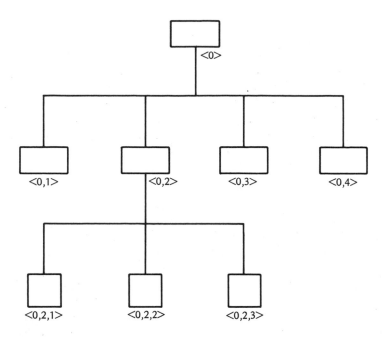

Figure 3.3B

```
WHILE  K% =  1
        GOSUB 01
        GOSUB 02
        GOSUB 03
        IF SIGMA  >  G  THEN  GOSUB  04
        ELSE
WEND
            K% =  K% +  1
PRINT  '*** PROGRAM  ESPBAS00  COMPLETED ***'
CLOSE
STOP
01  READ #02  J%;  SIGMA,  ALPHA$
    J% =  J% +  5: RETURN :
02  READ #01; TYPE$
    IF TYPE$  =  '     '  THEN  GOSUB  021 RETURN :
    IF TYPE$  =  'NICE' THEN  GOSUB  022 RETURN :
    IF TYPE$  =  'DARK' THEN  GOSUB  023 RETURN :
03  SUM =  STR(J%) +  SIGMA
    RETURN
```

```
04  DELTA = STR(J%) + SIGMA
    PRINT 'DELTA = ', DELTA
    RETURN
021 READ #01; DNAME$, DADDR%, DEMP, DSSN%
    PRINT DNAME$, DEMP, DSSN%
    IF DSSN% = 00000000 THEN 023  RETURN :
022 INPUT 'ENTER SOL SEC #',  DSSN%
    PRINT #01; DNAME$, DADDR%, DEMP, DSSN%, RETURN :
023 INPUT  'ENTER ANOTHER NAME';  ALPHA
    IF ALPHA = '        ' THEN K% = 1  RETURN :
```

The concept behind structured labelling is to *refine* the structure of control paths in a program. Whole numbers are used to label each refined function in such a way that the labels of progressively subordinate functions are concatenated with stepwise increasing digits. Functions at higher dimensions (lower for a tree structure) are concatenated with additional integral digits. The number of all the label digits of a function complex is called the dimension of its complex, where a complex is the box describing that function. Each label is unique. The structure label must reflect the hierarchical position of a complex. The recent developments in hierarchy structuring, called the trace of hierarchy, finally give powerful vindication to these claims:

The greatest value of structured labelling is this:

Given any function or data anywhere in the program, we can trace our way unambiguously back to the initial entry into the program, as well as down to the final exit from the program.

In other words, a program is not just a mass of structures of statements, but a directed graph. Every algorithm fits comfortably at some point in the hierarchy. Not only do the sequence structures have unique one-entry/one-exit, but each subordinate one-entry fits perfectly into a superordinate one-exit. Algol with its telescoping nests of BEGIN-END constructs lacks this structure label feature that would "piece" these nests together on a structure diagram. The most convenient realization of this labelling methodology is by means of numbers. The increasing pattern of numbers identifies the higher dimensions of the complexes. It is necessary and helpful to have these structure labels unique.

For those programming languages that allow it, structured labelling is the "perfect" attribute of a well-structured program. It serves as a means of not only obtaining a count but also controlling the paths of the functions at any level of abstraction. The discrete sequences of structure labels correspond with the distinct sequences of single-entry/single-exit structures.

3.4 STRUCTURED TESTING

---◆---

By structured testing we mean the following sequence of testing of computer system software:

1. Hierarchical testing
   a. Construct JCL (job control language) or WFL (work flow language)
   b. Construct dummy files
   c. Construct program stubs
   d. Test (execute) the JCL deck
2. Module testing (some modules may be tested independently)
   a. Path testing
   b. Function testing
   c. Data testing
3. System testing—integrates all modules for joint testing.

It would be questionable to combine (1) and (2), but the reason for the separation is more of convenience of perception and meaning than of methodology. The popular term "top-down testing" may have applied, but since this is inadequate for expressing dimensionality (top-down suggests vertically downwards), hierarchical testing as a basic testing concept serves to bring to focus that several modules may actually be involved at the same level of abstraction. Those modules that derive from the same node and are at the same level of abstraction will necessarily be tested at the same time—simultaneously, if possible, together with their node module.

Hierarchical testing, in effect, is more of an organizational judgment of how the constituent modules will be tested. Normally, this judgment is based on the nodes of the system's tree structure. Initially, the sets of job control languages for each module are constructed [34, 46].

The dummy files and program stubs are created to test the job control languages. Theoretically, this must be the first level of testing. The second level of testing begins by testing the nodes at the highest (first) level. Then, for each node at the second level of the tree, similar module testing is conducted at the third-level testing. We must note that each subsequent level of testing may involve multiple tests, depending on how many "independent" nodes are at the previous level. Each "independent" node and its subordinates lead to a single test. It is assumed each set of test modules at some level are tested together with their node module.

In summary, if a tree structure has N levels, there are at least $N + 1$ levels of testing. The additional level is for the job Control language tests, and each level may have more than one set of test modules. The total number of testings at any level depends on the number of superordinate nodes at the previous level. This is justification for including a hierarchical test level.

The second state of structured testing is module testing. As described above, this is in no way separate from or independent of the first stage. We treat it separately only for purposes of study. However, we include here the testing of those modules that are not dependent on others. The testings described for the second stage are therefore part of the testing described for each module at any of the levels of the hierarchical testing. For each module, the following tests are critical [25]:

1. *Path testing*—satisfies that each function path is efficiently and effectively traversed; given the right data, module execution terminates successfully.

2. *Function testing*—satisfies that every function structure is effectively computed using the correct parameters, and fails computation with the wrong parameters.

3. *Data testing*—satisfies that:

   a. Given the correct data and together with (1) and (2), the expected output data are satisfied and that the module executes conclusively and successfully.

   b. Given the incorrect data, the module fails, and the point of failure is unambiguously detectable.

The resulting difference between function testing and data testing is that the latter puts out the correct information expected of the program, while the former tests for the value acceptance of correct data. A practical proof for demonstrating that a path or function or even data element has been processed in a given module is to use a preformatted display at selected control points. The complete displays together give a fundamental structure of the program's logic. This device, called Datapath, is additionally instrumental in studying the control flow of the module; further, it has the capability of revealing flaws or other bugs in the program. It is particularly useful for testing modules whose structure is unknown. There are other advantages and extensions.

There are diverse debugging machines (tools) marketed by hardware manufacturers and software houses: dynamic dumps, dynamic monitors, traces, etc. These are "really just cosmetics" designed to resurrect a "dying program" [56, p. 117]. They are curative, not preventive. A well-structured module has less to no need for them.

3.5 MODULAR PROGRAMMING

◆

The melting pot of ideas that quickly kicked off the structured revolution is the modular revolution. Many of the philosophers and practices of modular programming also have their share of the confusion. Some centers even interpret a modular program to be equivalent to a single function. But how these functions are empirically defined is not discussed. Are they defined on decomposing or composite or irreducible minimal functions? Many therefore code the algorithm of a function, give it a label, add it to the systems library and call or copy it into subsequent "programs." There is no difference between algorithms, program or module. These partial compositions (of modules) are what Parnas calls "a responsibility assignment rather than a subprogram" [41].

An algorithm is a finite sequence of instructions; a program is a sequence of algorithms; and a module is a program together with all its necessary data structures (Figure 3.5A).

A rational philosophy of modularity found in a 1970 text for design of systems programs by Gauthier and Ponto [41] conceives modularity of systems as well-defined inputs and outputs without any confusion between any two "modules." Essentially, each task forms a separate, distinct program module. This enables testing and maintaining of a system in modular fashion in which bugs and failures can be easily traced to offending modules. This device therefore avoids indiscriminate searching for errors throughout the total system (of modules).

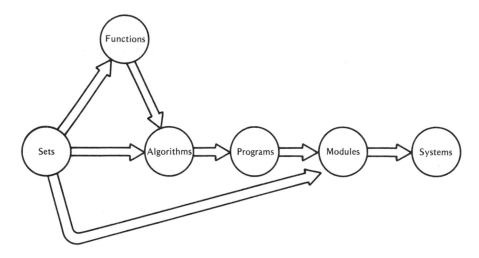

Figure 3.5A   Hierarchy of systems concepts in software engineering

Some of the benefits of modularity revolve around three major rea-sons. First, managerial convenience—a great deal can always be accom-plished over a shorter time by breaking complex ideas into simpler parts. Second, and again related to management, is the fact that changes are easier to apply to the parts individually than to the whole mass, particu-larly when such changes do not affect other components. This imposes a great constraint on the initial design of the modules. Third, with the modules defined according to optimality, it is easier to understand the total system by understanding the central goal of the individual compo-nents. It is like piecing together the parts of a machine.

It can be seen that these are the same goals of structured programming. But the key to understanding the concept, and therefore, definition of a module, is that described by Myers [39, p. 11] : a separately compilable and executable unit. This concept was used by the author to define a module as a mathematical system, called a *semigroup*.

In electronics and algebra where the concept of a module is well de-fined, a module is a construct over some ground structure; this is consis-tent with scientific thinking. Therefore, our ground structure will be the data structure, and the construct our program structure or set of al-gorithmic instructions over it. Thus, just as a system is more than an or-ganization, a module is more than a program. It is a program together with all its data structures, hence not just a function. According to the dictionary, a program is a description of a sequence of events or activities or instructions to be executed. This definition omits all the resources —manpower, equipment, etc.—required to effect those activities. This is consistent with the procedure division of Cobol language, although this contradicts the traditional use of the word *program* in programming. Consequently, the practice of defining a module to consist of mere algorithms is inadequate. The traditional "program" is the module, according to this semigroup definition.

Yourdon expects that structured programming will eventually be a basic acceptable method for obtaining modular programs. There is more to it. Modularity, like a feedback current, is an aid to structured programming. According to Myers [39, p. 11], a module is the primary unit of program structure. The most controversial question about modularity surrounds the criteria to be used in dividing systems into modules. Diverse propositions have been made and various approaches have been applied, ranging from number of lines of code to functions. Clearly, those definitions bordering on number of functions or lines of code or computer words are very inadequate, since we cannot demonstrate a logical and axiomatic justification for them. The methods of Parnas [41] as discouraging the use of flowcharts for decomposing are necessary. Their sufficiency raises the question of what is independence of a module, especially his conclusion of allowing subroutines and programs to be assembled "from various modules carries the same penalty as using functions or lines of code" as a definition of modules.

The paradox of the independence of a module stems from the difficulty of using the attribute of something to define that thing. There is obviously some bias. One purpose of modularity is independence (of execution from other modules). If we know where and how to define independence in a system, then the task of modular or structured programming would have been simplified. Unfortunately, independence is not an absolute property. In practice, some systems seem to defy independence of structure, while others do possess some nice clues to independence. In the latter, the data sets are clearly "distinct" for distinct "modules." It therefore follows that in order to solve the problem of independence three things are necessary in the system:

1. Data set conglomerates
2. Data structures
3. Function structures

Conglomerates are essential for defining the set of input/output structures. Therefore, to identify modules in a system, we employ the principle of inequivalence, as in chapter 5. The definition of a module must be free from arbitrary intervention.

The underlying cause for these conventional partial definitions can be seen in the prevailing rush to produce or an assumption that programming is like most ordinary human activities. On the contrary, the analysis, design and realization of a program is purely a mental activity—a Dijkstra human activity requiring thinking, planning and coordinating. Programming is different from construction of a highway—dig earth, pour concrete, pour water, cover with tar, paint boundary lines—because the context of programming is different. Concentration and mental effort are required to define, refine and assemble the isolated ingredients of a program. This necessitates the use of abstraction.

It is worth commenting here that the decision to use a modular approach to programming may also be considered a managerial tool, since it allows an organization to make an effective allocation of its personnel resources. However, such a decision is meaningful only after the modules have been defined.

EXERCISES

3.1 (a) Explore any other axioms for structured programming besides the six proposed here.

    (b) Call (a) the general set of axioms for structured programming. Select a minimal set of these that are satisfactory and nonredundant and call it the complete set of axioms for structured programming.

       Show that they are

         Consistent

         Necessary

         Sufficient

       Then if the above criteria are satisfied, the set will be called the canonical set of axioms for structured programming.

3.2   What are the contending factors for structured coding? What are the motivations for the revolution for improved (or structured) coding? Do you feel structured methodology has solved the problems?

3.3   Some programming languages allow labelling of instructions, addresses; others don't. Show the weaknesses and strengths of structured labelling.

3.4   Testing consumes considerable software development resources. Describe and document the differences between top-down and bottom-up testing strategies. Also criticize the concept of top-down. Given that a general theory of testing is not as yet well developed, define and describe some axioms for structured testing.

3.5   Is modular programming a point on the path for structured programming, or is structured programming a prologue for modular programming? Find a set of axioms on which to define a software module. Show that it is consistent and complete, necessary and sufficient.

Chapter 4

BENEFITS of STRUCTURED PROGRAMMING

Efficiency, elegance, performance and reliability are the four cornerstones of a satisfactory software product.

Poor reliability is disappointing, fatal and ultimately leads to total loss of confidence and, therefore, loss of revenue.

Poor performance is confusing, costly and can cause frequent maintenance, therefore draining economic resources.

Elegance, though considered trivial by many, is a manifestation of programmer creativity.

Efficiency, though considered secondary by some, is the sine qua non of a good design. In the functional architecture of an effective software product all four attributes must be harmoniously interweaved.

4.1 TO THE PROGRAMMER, DESIGNER AND ANALYST

The immediate benefits of structured programming are psychologically shared by those directly involved in its development and realization (coding, compiling, testing and debugging). Those who have applied the principles admit accelerated performance, professional satisfaction and excitement at good programming. There are few midnight calls about failed systems, and when such calls come, the nature of the failure will usually suggest to the programmer the point or neighborhood in the module where the bug can be found.

Above all, the existence of a set of principles which the professional may use to construct the building blocks of the system adds zest to his or her confidence, and vigor to his or her performance.

59

4.1.1. ACCELERATED PERFORMANCE

Structured programming is an educational challenge to those who have been in the field for sometime, and an intellectual initiation for beginners. The new student gains confidence in the tasks to be undertaken, having mastered a logical procedure for solving problems. The older student gains more proficiency after having been exposed to a scientifically manageable technique for handling computer systems.

One of the benefits of this education is improved desk-checking techniques. The flow diagram of Böhm and Jacopini and the hierarchical function operations are more manageable, traceable and understandable in their isolated subunits. The new coding technique, indentation, makes the program easy to read, and therefore more tractable. The direct result of these is increased programmer speed and accuracy.

Bugs become easier to trace and locate. In fact, most well structured programs have no use for traditionally cumbersome debugging utilities, and, consequently, testing takes less time than in spaghetti counterparts. In practice, the bugs in structured programs are less powerful than those in spaghetti systems and are always easily removable. They are more likely to be logical rather than structural errors. *Logical errors* are mainly those due to Boolean constructs, path predicates, coding errors and (we conclude) errors due to data type. *Structural errors*, on the other hand, are those due to analysis and design havocs of requirements definition and the like. These are technically more excruciating to correct. They may in certain cases cause the total scrapping and redesigning of the system. Since according to Dijkstra, the quality of a programmer is inversely proportional to the density of GOTOs in the program, the absence of GOTOs will directly increase the programmer's efficiency.

Finally, structured programming broadens the programmer's attitude toward systems. The scientific technique postulates a "general set of principles" for designing systems. Armed with these principles, the technician is freed from static ready-made answers to systems problems. He or she is open to questions and is ready to abstract new solutions for new problems, rather than trying to fit old habits and solutions to possibly more trying and complex new problems.

Above all, the programmer learns management of complexity. He or she learns to sort minuteness out of disorganization, to bring order out of anarchy. The mental capability to organize is a necessary practical step toward total management of resources (people and materials). Improved management ability leads to greater efficiency.

The point here is that management skill is a priori a self-discipline. It is an unconsciously accumulated reserve of experiences from observations, realizations, abstractions and analysis; these are the tools for management of complexity. The acme of management is manifested in quality decisions. We can teach people to manage things, but we cannot teach them to make decisions.

4.1.2. PROFESSIONAL SATISFACTION

Professional pride is usually associated with some type of education, either academic or in-service training. But more than professional pride is professional competence which reinforces professional pride.

Some schools (including on-the-job training) or seminars award special certificates of participation at the conclusion of a course as a reminder of the new accomplishment. However, of greater significance are the practical accomplishments to be made at undertaking new challenges and, in particular, at furthering the lessons learned by way of dedication, experimentation and/or contribution. This kind of professional satisfaction is a lifetime reward.

Another source of professional gratification comes directly from satisfactory completion of a task, such as with a successful walk-through or project review. The feeling of having satisfied the design or programming of a certain algorithm or module to general acceptability gives the professional a sense of pride and of relief from exertion. The thought of attaining this goal is instrumental for an adequate preparation (which is an intellectual activity) for the occasion. The professional realizes that his or her promotion will be reflected in his or her performance. In the absence of a satisfactory performance, the professional may learn from failure, which is not less intellectually challenging or awakening but fortifies him or her against future challenges.

A common anomaly in the computing community is meeting deadlines. The professional who constantly falls victim to this requirement usually is considered less useful and inspiring to the team. It follows that any method that improves the capability of meeting deadlines will instill professional pride in the individual. This is one of the objectives of applying structured programming.

4.1.3. DELIGHTFUL PROGRAMMING

Simplicity of thought, of conduct and of understanding are the basic requirements of an initiate. Simplicity of design, of coding and of testing are among the primary goals of structured programming. The popular adage in programming circles is "Keep it simple."

With these principles thoroughly mastered, the student gains greater insight into and expanded delight about programming. Programming ceases to be a boredom at the same time that it ceases to be a craft for superprogrammers only. Instead, it becomes applied science with known and provable techniques which greatly reduce the ordeal of debugging. Its place is taken by the intellectual creative exercises of designing and analyzing; the saved time can be used in other productive ways.

Added to these is the intellectual appetite to gain more programming style. The studious programmer begins to develop mastery of his or her "art," and thus researches new horizons. When the programmer can read his or her own codes as well as have another comprehend them, when the programmer can maintain his or her older programs as well as have a different programmer maintain them without considerable setback, and when the programmer can test and debug his or her own programs, he or she has reached the peak of the delight in his or her programming career. The initial orientation to structured programming, like learning a new language, is not easy, but mastery of the techniques brings new vistas of thought.

4.2 TO THE ORGANIZATION, CLIENT AND USER

―――――♦―――――

While the primary goal of structured programming was directed to improving program correctness and thus making the programmer benefit from the consequent simplicity, the same goal turned out to be an even greater benefit to the organization in three main areas: software productivity, software reliability and software maintainability. The experience and result from IBM's experiment with the combined managerial and technical aspects of the structured revolution have confirmed and popularized the benefits of structured programming, not only to individual programmers but also to the organization.

The efficient and effective management of structured programming is an ideal that should be aspired to by every computing environment. The techniques of chapter 2 combined with those of chapters 3 and 5 lead to considerable savings in the cost of software development and maintenance. Gone are the days of laissez faire developments and ad hoc flowcharting with their consequential frustrations and skyrocketing costs.

4.2.1 SOFTWARE PRODUCTIVITY

One of the most nagging problems in computer installations is the satisfying of production goals, in particular, deadlines. Experience with the IBM and New York Times project show considerable proof of having met the goals by using structured programming techniques. But more important than meeting the goals is the greatly reduced occurrence of bugs at acceptance testing. This further indicates that the systems would be less prone to sporadic maintenance. Even when maintenance phases were to be undertaken, still less time would be required for debugging. Baker [3] attributes a greater percentage of the success to the use of the chief programmer team approach, thus emphasizing the need for improved management. It was realized even while the project was still progressing that the technique was effective. This confidence has encouraged use of these new structured techniques in other divisions of IBM, as well as several other institutions and corporations. More remarkably, the software was delivered as scheduled despite 1,200 formal changes in the requirements, as well as cuts in manpower and computer budget.

Naturally, what is of utmost concern to the organization is not the style or the tool but direct savings in cost and increased profits. Obviously, the dramatic decrease in testing systems coupled with timely or even early delivery directly reflects increased productivity. It may turn out the organization will need less staff to undertake its systems goals, thus saving in personnel. The impact of reduction in staff also directly leads to reduction in material resources, since each staff member will need additional work resources. Both reductions cut waste and save operating revenue, which may be invested in more profitable ventures.

Associated with this is the growing consciousness about measuring programmer productivity. With programming becoming a science, this is a natural consequence. Part Two of this book presents some quantitative measures on programmer productivity, and it is hoped this will be a major contribution and inspire better metrics than now exist.

4.2.2 SOFTWARE RELIABILITY

Software reliability is a very critical factor for such systems as air traffic control, defense systems, medical equipment and medical diagnosis, bank accounts, real-time systems (e.g., laboratory data computer systems), and so on. Considering the heavy investments made in these life-and-death operations and the number of human lives affected, it follows that a near "ideal" reliability is required.

Software reliability is a measure of the probable presence of errors or bugs in the software (see Part Two). Reliable software is not necessarily one that is totally error free or proved correct. An organization is concerned with a measure of the relative number of detectible errors that can adversely affect operations within the software. Some bugs are not easily detectible for some time; as Dijkstra points out, it is virtually impossible to test a software exhaustively, for program testing can only show the presence, not the absence, of errors.

While reliability must not be confused with *correctness*, reliability may be synonymous with consistency and is related to intregrity. A reliable software need not be correct. Reliability is a concept from engineering, which borrowed it from the mathematical theory of probability. The difficulty with applying this measure successfully to software engineering stems from Dijkstra's observation of the impossibility of an exhaustive testing, compounded by a lack of knowledge of how to predict effectively the presence of errors as well as how to classify bugs. The latter follows from the nature of software engineering. A method to obtain a numerical measure is presented in Part Two.

Heuristically, reliability indicates that a software will or will not operate longer and satisfactorily before failure, or may not even fail (ideal programmer). The organization therefore has considerable confidence in a reliable software.

With these arguments, it follows logically that a well-designed, well-coded and well-tested software has a higher likelihood for better reliability and vice versa.

4.2.3. SOFTWARE MAINTAINABILITY

The three factors of benefit—performance, productivity and reliability—are all reflected in the frequency and efficiency of maintainability. In essence, maintainability is positively correlated with each of these factors, using the language of graphs and statistics.

Effective walk-throughs and project reviews have the result of reducing and detecting bugs long before testing is initiated, thus ensuring better reliability. These then lead to less sporadic maintenance, traditionally the woe of many professionals and the black hole of software budgets.

Much literature with graphs, sketches and other illustrations has been published about the skyrocketing cost of software maintenance. The curious reader must have realized that disciplined programming techniques can greatly reduce bugs, and hence subsequent maintenance. Better software reliability means maintenance will be easier and less time consuming and, consequently, less costly. Maintenance will generally be more of modification, revision and updating, and less of repair and patch-up.

However, if need arose for major maintenance, called enhancement, the hierarchical design structure and accompanying structured documentation will be of invaluable help. This follows from the fact that structured programs (software) have better design, in addition to better documentation, better coding and are better tested than their spaghetti counterparts, and therefore yield better maintainability.

Besides, the use of structured programming will bury the notion now current that maintenance is a chore nobody wants. On the contrary, experience has shown that structured programming for "structured" maintenance is equally rewarding and gives satisfactory results. Above all, one may have something to learn from maintaining another's software. The prevailing bias against maintenance was born out of a lack of principles for programming. The use of flowcharting was completely inadequate for tracing backward the origin of data items, in addition to being a time-consuming effort. Function charts, as used in structured documentation, unequivocally far supersede these unstructured techniques and are a great leap for maintenance.

It must be realized that the problem of readability is also very important for maintainability. A poorly coded structure can hide punctuation marks and other salient features, and thus obfuscate maintenance. Perhaps here is the best place in the book to include a remark on the so-called coded documentation seen inside programs. These are actually remarks. Assembly language programmers have this habit strongly entrenched in their practices, which is understandable since this language is "purely" symbolic, at least meaningful to the average programmer. But the excessive use of remarks in higher level languages is questionable. It has the capability to degrade maintenance. These ills of remarks can be eliminated by good structured programming, meaningful structure labelling and helpful data-naming techniques.

4.3 TO THE OBSERVER, ADMIRER AND CRITIC

———————◆———————

Dijkstra advocated a structured approach for two purposes:

1. Efficiency and elegance in programs
2. Proving program correctness—for what program structures this can be done, not how to prove them

The proof of program correctness need not be made by the programmer alone. Some researchers are devising axioms and predicate clauses for proving program correctness mechanically.

One cogent reason for advocating structured coding is readability. The difficulty of having one programmer or even the project manager read another's or a subordinate's code has obvious discouragements. A structure coded program has improved readability.

Part of the difficulty of having one person read and debug another's program stems from data-naming conventions. Structured data names must in some way reflect the meaning of data. A simple technique is to prefix every data of a record with some prefix of the file name.

4.3.1. EFFICIENCY, ELEGANCE AND PERFORMANCE

Efficiency in technology is the ratio of output of a machine (possibly human) to input. Elegance, in general, is associated with style, design and content. Performance is the effectiveness of managing the resources of the system toward satisfying the defined goal of the software. Performance satisfies that the software meets all timing and storage estimations. Choice of algorithms is vital. It follows that both efficiency and elegance in software must be applied from the design phase through coding and testing. The type of code produced at compile time contributes to the general efficiency; elegance has more to do with visual appearance of codes or design. A straightforward example comes from nested-IFs. The author knows one installation where the operating standards forbid nested-IFs. This is an invitation to inefficiency.

It must be admitted that these factors are directly influenced by the quality of designs. At the design phase also, the data structure has a conspicuous influence on efficiency. Some programmers feel that the large size of fourth-generation computers defies consideration of efficiency in design. Some Cobol programmers use very long data-names.

A sparsely described data structure or one with numerous pointers obviously executes slower and less efficiently than one with contiguous packing or fewer pointers. High efficiency has a direct influence on reducing operating costs. The factors for consideration of efficiency are time of execution, space and cost. Storage space and time are compiler and execution factors. Efficiency, therefore, is a measure of programmer productivity, and of programmer creativity—a necessary incentive to programmer productivity. The higher the efficiency of a program, the more advanced or sophisticated the programmer. Since compilers that produce efficient code are large and slow because of the extra processing optimizing code, it follows that efficiency should be a major concern throughout the life cycle, and beyond to maintenance. Many present-day software engineers have a tendency to ignore efficiency, particularly for smaller systems; this is simply an abuse "practiced by penny-wise and pound-foolish" programmers, a justification for the inability to maintain their own "optimized" programs. In any established engineering discipline, a 12% improvement in efficiency is considered very significant; the same attitude is worthy of attention by software engineers [56, p. 60]. The same view is also expressed by Denning—that "considerations of efficiency" are better initiated at the inception of the system's life cycle [56, p. 16].

There is an argument that structured programming is inefficient because of frequent calls to subroutines. This argument may be justified in that some programmers have a tendency to design too many CALLS to subroutines. This is inefficient design, particularly when such subroutines contain several procedures or many large functions. It is here that memory and access time are consumed—too many unwieldy parameters are passed between subroutines, and some of these subroutines even make calls to some other subroutines, leading to chains of calls. This situation is worse for external subroutines. Based on this, the complaint is genuine and justifiable.

On the other hand, if elegance in a solution is interpreted to mean simplicity, then elegance is something relative—only to be revealed in the natural properties of the system. In this case, a simplified design and coding leads to the most elegant solution. Simplicity then is the tendency to derive results with given or known conditions. These anomalies and more are the targets to which structured programming is directed to cure and correct.

These and other principles presented in this book do not in any way exhaust the wealth of principles of structured programming. The individual programmer has more to learn by gaining experience through exposure. This is the reason for the emphasis on programmer creativity for programmer productivity. Creativity is a derivative of personal experience, direct or indirect.

The following portion of a simple Fortran program is a typical example of degradation of performance (on virtual storage):

```
DO 16  K = 1,512
     DO J = 1,20
16        XZERO(K,J) = 0.0
```

Since Fortran arrays are stored by columns, each column requiring a 4K page, each execution of statement-label 16 will reference a different page. The inner loop will execute 20 pages for a total of 512 times. In order to improve performance, it is advisable to reverse the order of the loop so that K = 1, 512 becomes the inner loop. This will necessarily decrease cost by a ratio of 512 to 1 [46].

Another example of speed of execution contrasts the two examples below. Clearly, the first example executes faster because of efficient structuring of the nested-IFs; actually, there is only *one* statement.

EXAMPLE 4.3.2A: Structured Nesting

```
20-STRUCTURED-NESTED-IF.
    IF   cond-A
        IF   cond-B
            IF   cond-C
                DO   210-NESTED-EXAMPLE-A
            ELSE
                DO   220-NESTED-EXAMPLE-B
        ELSE
            DO   230-NESTED-EXAMPLE-C
    ELSE
    IF   cond-D
        DO   240-NESTED-EXAMPLE-D
    ELSE
        DO   250-NESTED-EXAMPLE-E.
```

EXAMPLE 4.3.2B:

```
20-UNSTRUCTURED-NESTED-IF.
    IF  NOT  cond-A
        DO   250-NESTED-EXAMPLE-E.
```

```
IF   NOT   cond-B
     DO   230-NESTED-EXAMPLE-C.
IF   cond-C.
     DO   210-NESTED-EXAMPLE-A.
ELSE
     DO   220-NESTED-EXAMPLE-B.
IF   cond-D
     DO   240-NESTED-EXAMPLE-D.
```

Other significant issues of efficiency, elegance and performance are the choice and design as well as the refinement of algorithms. This is critical for scientific and engineering systems, however it cannot be ignored for commercial systems. Efficiency in the design of algorithms reflects the programmer's education and competence.

Efficiently designed algorithms execute faster and utilize compact storage memory, thus improving computer performance. The efficient manipulation of algorithms is one of the cogent reasons for advocating functional thinking in structured programming. Once a function is defined, all that changes is its inputs.

EXAMPLE 4.3.2C: Use of the INDEXED BY clause of Cobol is more efficient and faster than the use of a SUBSCRIPT to manipulate a table. The former is a relative addressing pointer, while the latter is a counting device requiring the operations of incrementing and pointing (to the corresponding address).

EXAMPLE 4.3.2D: In the (hierarchical) design of data structures, it is more efficient to assign a group name to a contiguous set of elementary data items. This minimizes coding and speeds up execution by referencing a single address rather than several. These group structures may not have been identified at design time but have to be perceived and amended by the vigilant programmer at coding time.

4.3.2. PROVING PROGRAM CORRECTNESS

A correct module is one that satisfies its requirements definition. Module correctness should be valid for certain classes of module structures. There is no established rule—theoretical or empirical—for obtaining an estimate of the number of errors in a system. In other words, there is no proof of absolute correctness. Testing can, at best, detect the presence of bugs and not their absence; moreover, exhaustive testing is an unrealizable goal. In a limited way, structured (or organized) testing can improve the degree of correctness and the confidence of users. Consequently, degree of correctness falls into the class of probability measures. On the contrary, performance measures are taken to be jointly dependent on storage size and execution time and are therefore application and machine dependent.

In addition to efficiency and elegance, two other requirements of any software system are correctness and performance, which can be accomplished by either of two ways [43]:

1. Active or theoretical—takes the form of axiomatic program proofs

2. Passive or practical—takes the form of traditional testing and debugging

The purpose of testing is to determine the degree of correctness and reliability. If a program has no errors, any set of test predicates may prove reliability since all bad data will be kicked out. However, the problem of correctness is by far too gigantic to be manageable, least of all measurable. Program correctness can only be judged from the following levels:

1. Requirements definitions—how well the test or proof satisfies the requirements

2. Design specification—how consistently the test or proof validates the design

3. Implementation definitions

   a. Coding—how well the test or proof justifies the coding and hence how well the coding effectively conveys the ideal of the design, therefore achieving the goal of the requirements definition

   b. Testing—how efficiently the choice of testing strategy (and therefore test data) demonstrates the presence or absence of bugs and thus attests to the program's reliability and performance.

These and other latent conditions make the active approach very costly and impractical; the passive approach is time-consuming (therefore, costly) and infinitely inexhaustible and formidable. Despite this, a modest, organized choice of testing strategy appears to be the best alternative. The theoretical approach has one obstacle that makes it uneconomical (timewise and costwise), namely, the diversity of variability of input data and types. The mechanical proof, even for a small program, has the capability to become voluminous. Moreover, there does not seem to have been developed any well-known set of axioms. Until such a time, theoretical or mechanical proof of program correctness will remain more of a fiction, perhaps to await some structured testing revolution. There is no doubt that structured programming, as opposed to the traditional anarchy of spaghetti systems, can and does lead to better correctness proofs. With its manageable measures, structured programming does much to stimulate better theoretical and practical approaches.

This follows because program correctness is meaningful if and when we have a well-structured system or module (WSS or WSM); there is no known way to measure anarchy. The first concern of program correctness proof is the degree of structuredness. If, for instance, a module has the EXIT-IN-THE-MIDDLE structure, its correctness proof becomes more difficult because of its degraded degree of structuredness. For a meaningful proof of correctness, we must be able to trace an arbitrary data element from an arbitrary source to a corresponding destination. This makes use of the practice of structured labelling or any equivalent (in any programming language). This is what is called path predicate proof.

The second concern is degree of complexity. A very complex software naturally tends to frustrate correctness proofs—active or passive. In the active approach, the theorems and llemas increase positively exponentially; in the passive approach, the required test data may become voluminous. In either case, the path predicates also add to the complications.

The following conditions for testing software justify the claim to the complexity of correctness proofs:

1. Path predicates—testing every branch, including selection conditions and all appropriate data

2. Predicate variables—testing the variables that determine the path predicates

3. Function variables—testing the correctness of functions and algorithms

4. Termination criteria—testing every parameter for a terminating condition

5. Miscellaneous—data integrity, validity, etc.

However, the task of correctness of proof testing may be simplified by considering a software system as a composite of functions, and thus concentrating on these functions individually according to their hierarchical occurrence.

EXERCISES

4.1 Tabulate the advantages of structured programming to the programmer. Describe any weaknesses.

4.2 Repeat 4.1 in the case of the organization.

4.3 Repeat 4.1 in the case of the critic or skeptic.

4.4 Use the arguments of this chapter to attempt a justification of the arguments against structured programming, including those mentioned in chapter 1.

4.5 Use the proposition of this chapter to motivate the development of theory for structured programming.

Chapter 5

PRINCIPLES of STRUCTURED PROGRAMMING

"Abstraction is the realization of reality."

Dr. Steve J. Bryant
California State
University, 1969

The process of realization can be enhanced by symbolizing, sketching, graphing, mapping, computing and differentiating or integrating. Reality is homogeneous, but the realization of reality is heterogeneous. Any system may be realized in terms of its inputs and outputs.

The process of abstraction consists of isolating the components of the things realized and then concentrating on their relationships with respect to each other in order to construct the functional architecture of that which is realized.

5.1 FUNCTIONAL BASIS FOR STRUCTURED PROGRAMMING

———————◆———————

The concept of basis seems to have been well grounded in the developments of mathematics, in particular linear algebra. Briefly, the concept of basis seeks to establish a non-empty set of structures or building blocks upon which all other structures of that concept or system can depend.

In other words, given a structure belonging to that concept or system, we can demonstrate that it can be made up or built up from the elements of the accepted basis. As a familiar analogy, a house is built on pillars as a set of basis.

This idea of basis may sound insignificant at first reading, but it is the very precious ingredient for which we have unconsciously been searching toward understanding programming. In geometry, where the notion of basis has its origins, lines that intersect at right angles (orthonormal) are taken as bases and every point in that plane is referenced to these lines. In advanced algebra, the same concept is modified in terms of equations of these lines; and in group theory or abstract systems, we still refine the idea in terms of generators of elements. The argument here is that we can advance an analogous notion for understanding structured programming. Perhaps that is the meaning of *structured*.

We shall explore the practical applications of this notion in the sections that follow. In this section we want to reinforce the emphasis on using function thinking and to add that a function is not necessarily always as complicated as it appears in mathematics. Human thinking is generally function oriented; many of the verbs of natural languages are "function" expressions. Finally, we want to establish that there does exist a generalized (scientific) methodology for designing systems and programs. The theory of structured programming postulates that a reliable solution to every problem must gravitate around the natural proclivities of that problem.

In scientific programming languages (Fortran, APL), functions are explicitly defined. In commercial programming languages (Cobol, Snobol), on the other hand, functions are generally implicitly defined and sometimes vaguely. However, they can always be reformulated to make them amenable for computer processing. The advantage of functional thinking is that each function explicitly "knows" its input and output (data structures). Thus the "boundary" between any two functions is easy to draw.

EXAMPLE 5.1A: Explicitly Defined Function

An explicit function essentially specifies *an* algorithm: what to do with input (in the domain) to obtain output (of the range).

$$f(x) = x^2 + 4x$$
At $x = 0$ in the domain, $f(0) = 0$ in the range
At $x = \frac{1}{2}$ in the domain, $f(\frac{1}{2}) = \frac{1}{4} + \frac{4}{2} = \frac{1}{4} + 2 = 2\frac{1}{4}$ in the range.

EXAMPLE 5.1B: Implicitly Defined Function

Update Customer file: The update "function" is one of a class of partial functions. Some record types may not qualify; some field names of those qualifying record types may not be updated, etc. In general, we write $U:R \rightarrow G; R$ = set of records on file; G = set of updated records. Thus

$$RU = (R_1, R_2, \cdots R_H); (U_1, U_2, \cdots U_k)$$
$$= R_i U_j = \begin{cases} G_i & \text{if } i = j \\ \phi & \text{if } i \neq j, = \text{No Action (on this record)} \end{cases}$$

A function is a *distinct composite unit of thought*: distinct because its relationship to other things (including functions) can be clearly delineated, and composite because its contents (inputs and outputs) are implied in its definition. A function generally has a name by which it is addressed; this name is either explicitly stated or implicit in its notion. The value of functions for proving program correctness can never be understated. A function, once defined, lends great ease to describing algorithms. Its form remains fixed but its data may change in time. Given a formal or implicit function specification, all its inputs and outputs can be constructed, thus simplifying the correctness proof that the described algorithm does or does not satisfy the original requirement or specification. This special property of functions is inconceivable and impractical for data-driven or flowcharted systems.

A major handicap to this functional approach is the habitual cynicism that functions belong to the domain of pure and applied mathematics only. The verbs (commands) of everyday language are potential functions; anything that has inputs and outputs is a description of a function, however unsophisticated. Consequently, the functional approach will help reduce programming bugs and directly improve program correctness and reliability. This is because the programmer can concentrate on those functions individually during software design as well as during coding or testing. In practice, these functions can be seen as the nodes of a tree diagram where each function is the immediate root of all subfunctions that have resulted directly from its decomposition (whenever this is possible).

Finally, a practical justification of the functional approach is the concept of *partial functions* which dominate most of the functions of information processing. Given a global (data) set, it is possible to partition it into subcomponent sets. Then any function defined on the global set is partial if its total domain is one or some but not all of the subcomponents of the global (data) set. In other words, the (partial) function has meaning to some elements of the global set and not to others. On the contrary, if a function is defined over every element of its domain, that function is said to be total.

EXAMPLE 5.1C:

A global set is comprised of the subdomains of the set of Latin alphabets and decimal digits. The (partial) function ADD is definable over the subdomain of decimal digits, but is meaningless (conventionally) over the subdomain of Latin alphabets. On the other hand, the function CONCATENATE is a total function on the global data set.

5.2 FUNCTIONAL BASIS FOR RECURSIVE PROGRAMMING

———————◆———————

A common misconception among programmers, particularly commercial programmers, is that certain "new" ideas belong strictly to academic environments, and have no value in commercial applications. The general remark is "we don't need it." This is one of the reasons the structured revolution was delayed in catching up with several programming communities. In reality, some of these ideas may have had their roots in elementary mathematics (high school level). Perhaps the reluctance to accept these ideas is justified by academic linguistic clothing of those notions. Functions fall within this category, and worse than this is the class of functions called recursive functions. In its most basic meaning, all functions that are computable in any intuitive sense are said to be recursive. Hence, the functions SUM, PRODUCT and POWER are recursive [6, p. 20].

EXAMPLE 5.2A:

The most common and simplest example of a recursive function is the factorial function. By definition, the factorial of a number K is written

$$K! = K \cdot (K - 1) \cdot (K - 2) \cdots (K - (K - 1))$$
where $\ 0! = 1$, by definition. Thus
$$3! = 3 \cdot 2 \cdot 1 = 6$$
$$4! = 4 \cdot 3 \cdot 2 \cdot 1 = 24$$

The characteristic notion of recursion is self-reference. Those functions that are defined inductively have the character to use earlier arguments to compute current values. However, not all recursive functions are defined inductively. For those that are, the controlling mechanism is induction.

EXAMPLE 5.2B: Inductively Recursive

Define $F(0) = 0$; $F(1) = 1$ and $F(N) = F(N-1) + F(N-2)$, $N > 1$. The first few members of F are 0, 1, 1, 2, 3, 5, 8, 13, 21, 34, ... F is the famous Fibbonacci sequence of numbers.

EXAMPLE 5.2C: Non-inductively Recursive

```
If time is less than one year
    PRINCIPAL = PRINCIPAL
ELSE
    PRINCIPAL = PRINCIPAL + PRINCIPAL X TIME X INTEREST
```

In general, the process of recursion enables us to initiate and complete an activity by successive stepwise simplification of the activity up to termination. Recursion is an important concept in programming, having its roots in elementary mathematics. As a programming technique, recursion conforms to the top-down concept [34, p. 137]. In order to solve a given problem P, say, we should investigate the possibility of obtaining progressively partial solutions in such a fashion that one solution at each step simplified the problem. Then, proceeding in this telescoping order, we finally arrive at the ultimate solution of the initial problem. In essence, at each step we define "another" problem but one of lesser complexity than before. Each newly defined problem-step may invoke the "resources" of any prior solution. There may even be a chain of invocations.

However, this process must ultimately terminate. To guarantee termination, we must provide that any chain of invocations must eventually be sufficiently simple to complete the solution. This is the task of the designer.

The problem of termination reveals the inductive nature of recursion. Whether disguised in the use of switches or numerical count, the property of induction provides for ultimate termination.

In applications, certain programming languages (notably Fortran and Cobol) prohibit recursive invocation of programs [55, p. 87]. In a language with static memory allocation, the activation record (actually a stack) of the called module may be lost if a module calls itself. In contrast, those languages that provide for dynamic memory allocation (notably Algol, PL/1) afford a new activation record at invocation.

EXAMPLE 5.2D: Recursion Specification

```
DECOMP      PROCEDURE              RECURSIVE
            IF  PROB  (P)  IS  MINIMAL
                CALL   FINISH  (P)
            ELSE
                BEGIN
                    DECOMPOSE  PROB  (P)
                    CALL   DECOMP;
                END
        END  DECOMP;
```

The Cobol programmer may recognize the above example translated into the (iterative) form:

```
OBTAIN-SOLUTION-P.
   SET PROBLEM-COND-SW.
   PERFORM SOLVE-PROBLEM-P
       UNTIL PROBLEM-COND-RESET.

SOLVE-PROBLEM-P.
      IF Problem-P is sufficiently simple
          PERFORM COMPLETE-PROBLEM-P
      ELSE
          PERFORM REDUCE-PROBLEM-P.
REDUCE-PROBLEM-P.
        STATEMENT-1.
        STATEMENT-2.

        . . . . . . .
        . . . . . . .
        STATEMENT-N.
COMPLETE-PROBLEM-P.
        STATEMENT-1

        . . . . . . .
        . . . . . . .
        STATEMENT-N
        RESET PROBLEM-COND-SW.
```

As another illustration, consider the program complex UPDATE. Depending on the data structure and the function specification, this program complex may require partial updating. We may have to decompose the function in such a way that we obtain three or more partial subfunctions to accomplish our goal. This is analogous to the technique of partial differentiation—a technique of divide and rule. This example strengthens our argument that there do already exist known techniques for solving general systems problems.

Several examples of recursion come from operations research—under the title of *dynamic programming* or *recursive optimization*—where problems of allocation of resources are critical. In them the ultimate solution is arrived at by dividing the problem into *stages* and concentrating on one stage at a time. In practice, the solution of each stage is to transform the elements from one state to another state which then begins the next stage. A good example is the famous stagecoach problem or the travelling salesman.*

In conclusion, recursion and iteration have the common property that the terminating conditions (parameters) must be explicitly and properly defined. Both use inductive processes. However, they differ as follows:

Recursion:

1. *Self-addressing*—the function being computed may call itself.

2. *Dynamic*—implicit top-down-like computation, i.e., computation has telescoping levels to termination as a simplification process.

3. *Initialization*—Initializing addresses may be implicit or defined by function.

Iteration

1. *Non-self-addressing*—control calls to routine are externally defined; terminating checks are made at each inductive step.

2. *Static*—single-level computation, i.e., computing function does not establish any simplifying levels; the last inductive step contains the current data.

3. *Initialization*—absolutely essential before entering computing routine.

---

*The reader interested in these and several other dynamic programming problems may consult [26].

Finally, the overriding decision to use either recursion or iteration is performance (storage size) and clarity. Naturally, computing more complex functions may benefit from recursion.

5.3 CANONICAL BASIS FOR STRUCTURED PROGRAMMING

———————◆———————

The term canonical as used here means *accepted*. No one disagrees that it was Dijkstra who introduced the concept of hierarchy, or that Mills introduced the extension known as single-entry/single-exit [36], or that Böhm and Jacopini proved the famous theorem that bears their names.

The foregoing chapters have unfolded the story of structured programming. This section will treat the fundamental principles of structured programming. These are the basic principles around which all the applications of chapters 2 and 3 revolve. Their consequences are the developments of Part Two of this book.

In its strictest sense, the question, "What is structured programming?" has its answer in the pure applications of these basic principles. Any logical philosophy must have a set of principles as its pillars (basis). The Böhm-Jacopini basis element is the *syntax*—the what; and the Dijkstra-Mills basis element is the *semantics*—the how of structured programming.

5.3.1 FUNDAMENTAL STRUCTURES— THE BÖHM-JACOPINI BASES

Using the concepts of flow diagrams (function) and their decomposition, Böhm and Jacopini, in a landmark paper entitled "Flow Diagrams, Turning Machines and Languages with only Two Formation Rules" [4] laid down the theoretical basis for structured programming. This barrier between the theoretical possibility and the actual practicality of structured programming has not completely dispelled the disbelief of those "application only" professionals in the commercial processing environments.

The main result of the paper is that it is possible to describe any program completely by means of a formula containing the names of three fundamental structures (example below). We shall call these structures the Böhm-Jacopini Bases for structured programming. Each basis element will be presented here, but without reproducing the Böhm-Jacopini theorems.

EXAMPLE 5.31A: Three Classes of Instructions

There are three nonequivalent classes of instructions:

D_o = DO-ONCE; D_a = DO-AGAIN; and D_e = DO = PATH.

It follows that if S_i is an instruction (statement), then S_i belongs to exactly one of D_o, D_a or D_e, that is

$$S_i \epsilon \{[D_o], [D_a], [D_e]\}$$

For instance:

$(X = Y + 4) \epsilon D_o$
$(\text{Sum} = x + j \text{ for } j = 1, N) \epsilon D_a$
$(\text{IF } (K.NE.X) \ X = X + Y) \epsilon D_e$

First, let's introduce what will here be called a program or problem complex, or complex for short. A program complex will be any *portion* of a program or algorithm capable of being decomposed or refined up to minimality, or irreducibility, into elementary (function) structures. A complex is simply a cut made in a program provided it is or may be composed of sets of elementary sentences, and is representable as flow diagrams.

EXAMPLE 5.3.1B: A Complex

IF (a < ALPHA AND > BETA) SUM = SUM + ALPHA
ELSE
    CALL SIMPLEX(ALPHA, BETA, DELTA)

The whole complex belongs to D_e, but its components are as follows:

(SUM = SUM + ALPHA) ϵD_a
(CALL SIMPLEX(ALPHA, BETA, DELTA)) ϵD_o
(a < ALPHA AND BETA >) is a Boolean predicate of the D_e.

The notion of flow diagrams comes from graph theory where it is used to define a path for the flows of events or sets of elements obeying a certain rule. The use of flow diagrams has become customary since the advent of automatic computing. Flow diagrams are useful constructs for charting the flow of program instructions, preparatory to writing the actual instructions to be input into the computer. In its simplest form, a flow diagram has three basic parts:

1. *A process* (rectangular) box containing the pseudo-instructions to be executed.

2. *A decision* (predicate) box for selecting one of several paths of process boxes; it is traditionally represented in a diamond shape.

3. *An arrow* indicating direction of action (flow).

The three basic notions used in working out the Böhm-Jacopini theorem are:

1. Functional boxes

    Type I: Representing elementary operations (Figure 5.3.1A), e.g.:

           ADD A TO B
           ADD P TO Q

    Type II: Representing predicative decisions (Figure 5.3.1B), e.g.,

           IF A then B ELSE C

    Type III: Representing repetitive decisions (Figure 5.3.1C)

2. Mapping—Abstract representation of flow diagrams (for purposes of identifying equivalent flow diagrams).

3. Normalization of flow diagrams—a process of reducing or decomposing complex flow diagrams to their fundamental building structures.

Figure 5.3.1D is a flow diagram of a rather complex type. The question that Böhm and Jacopini ask is this: Is there a way to reduce this diagram to simpler structures? If so, can we always do it for any given flow diagram (of any program)? The answer to the first question is yes; the answer to the second is no—some diagrams are already in their minimal decomposed basis structures (Figure 5.3.1E).

Figure 5.3.1F shows one method of making this cut or decomposition. Figure 5.3.1G substitutes these decomposed structures to simplify the original diagram. And Figure 5.3.1H is a further simplification using the substituted structures. It is clear that this decomposition is not, in general, unique.

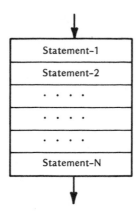

Figure 5.3.1A SEQUENCE or FUNCTIONAL box: Statement-1, State-ment-2, . . . , Statement-N form the SEQUENCE of in-structions to be executed (in that order).

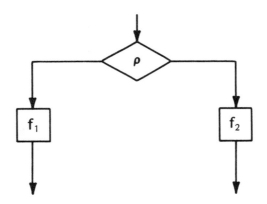

Figure 5.3.1B Predicate decision: p is the predicate; and f_i is the func-tion to be executed if p is true.

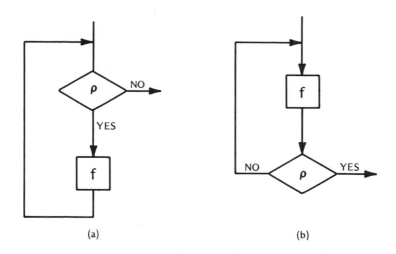

Figure 5.3.1C Iteration: (a) DO f WHILE P (is true); (b) DO f UNTIL p (is true).

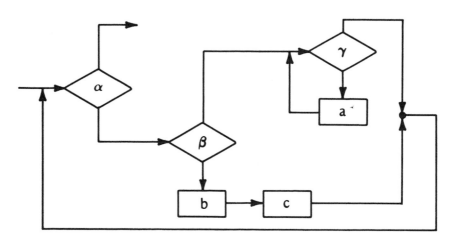

*Figure 5.3.1D A sample complex—adapted from [5].

*Adapted, with permission, from Böhm and Jacopini, "Flow Diagrams, Turing Machines and Languages With Only Two Formation Rules", *COMMUNICATIONS OF THE ACM*, Vol. 9, No. 5, Copyright 1966, Association for Computing Machinery, Inc.

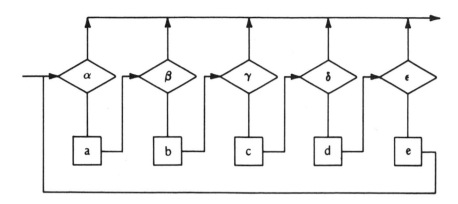

*Figure 5.3.1E IRREDUCIBLE COMPLEXES or MONADS: adapted from [5].

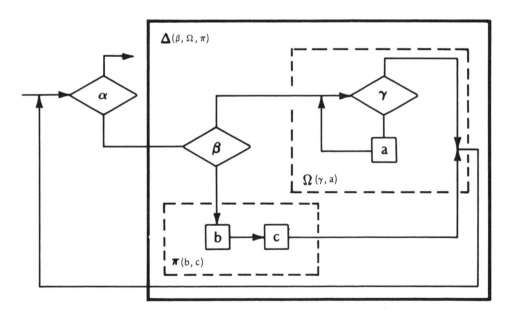

Figure 5.3.1F A sample cut in Figure 5.3.1D

---

*Adapted, with permission, from Böhm and Jacopini, "Flow Diagrams, Turing Machines and Languages With Only Two Formation Rules", *COMMUNICATIONS OF THE ACM*, Vol. 9, No. 5, Copyright 1966, Association for Computing Machinery, Inc.

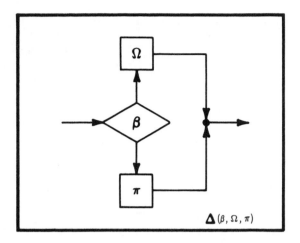

Figure 5.3.1G A symbolic substitution for Figure 5.3.1D

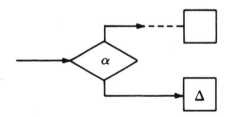

Figure 5.3.1H Simplified Figure 5.3.1D by substitution

The three bases structures of Böhm and Jacopini are as follows:

1. *Sequence box* (Figure 5.3.1A)—contains a sequence of statements which are executed in their order of arrangement or occurrence.
2. *Selection* (Figure 5.3.1B) or alternation—a decision structure that allows the choice of only one of two flow paths.
3. *Repetition* or looping (Figure 5.3.1C)—a sequence structure that is executed a given number of times.

For purposes of clarity and simplicity, we'll represent these structures with the following symbols:

1. Σ for sequence
2. Λ for selection
3. Π for repetition or looping

We define the Böhm-Jacopini bases as the totality of all three (program) structures. Each is a basis element. In other words, any program structure can be represented by one or more of only these bases elements of structured programming. Henceforth, we shall perceive program structures as complexes of composite functions capable of being refined step-by-step to one or more of these three (basis) structures.

Finally, let's generalize the selection structure. In practical applications, several criteria for selection may be encountered where each criterion corresponds to a particular path. In unstructured constructs of Cobol, such issues were handled by the DEPENDING ON clause. Fortran traditionally lacks the ELSE option of the SELECT statement. Our structured coding for such conditions is as follows:

```
IF   cond-A
     PERFORM   statement-A
ELSE
IF   cond-B
     PERFORM   statement-B
ELSE
IF   cond-C
     PERFORM   statement-C
```

```
ELSE
IF   cond-Y
     PERFORM   statement-Y
ELSE
     PERFORM   statement-Z.
```

The technique for handling this type of structure (Figure 5.3.1I) is through the GENERALIZED SELECT or CASE statement, which is not yet implemented in some IBM Cobol, unlike Burroughs. The construct is

```
Case:
i          Do Statement-i;
```

where i is a variable condition, e.g., cond-A, and ranges over some path variables. A variant of this exists in PL/1.

Finally, it is natural that one of the alternative paths of Figures 5.3.1B or 5.3.1I may further determine another choice. In this instance, we obtain the familiar nested-IF which is representable as a binary tree. The situation may repeat any number of times from either path. This graphical approach simultaneously reveals the hierarchical order of the nesting as an intrinsic property of the system. The conventional method of graphing is not admissible into a tree diagram, i.e., it has no room in a hierarchical diagram. This "newer" structure is much simpler structurally and functionally than the conventional diamond-structured nested-IFs. It fits perfectly as a node into a tree diagram.

But the critic may ask, Where is the single-entry/single-exit feature of this structure? This feature is a property of the hierarchical basis; it is not, by itself, a basis structure or element. It is a mere extension. The

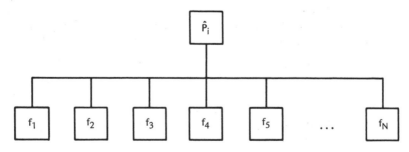

Figure 5.3.1I   Generalized SELECT or CASE structure: DO f_i IF p_i is true

Böhm-Jacopini bases elements are concerned with individual structure elements, and not with the order or method of accessing them. The two must *not* be confused—basis feature and use of a basis feature—or else we will be confusing the syntax and semantics of the "grammar of a language."

5.3.2 HIERARCHICAL STRUCTURES—THE DIJKSTRA-MILLS BASIS

The Böhm-Jacopini theorem completely and conclusively answered the question that any program can be written using a combination of the three fundamental structures discussed in the previous section. This enables us to analyze, design and code program structures (instructions) in terms of these fundamental structures. But the Böhm-Jacopini technique does not indicate any logical order of decomposing or refining these structures in order to maintain their group subordinate sequence. One solution to this appears to be the use of segmentation.

Segmentation connects one procedure or algorithm with another. But this other procedure may not be immediately subordinate, i.e., the single-exit of the first procedure may not fit immediately with the single-entry of the connecting procedure. We would like to have a structure where a global path's one-entry/one-exit local paths are sequentially strung together as "pearls on a necklace," as Dijkstra said [16].

The local paths, P_i connect adjacent nodes, N_i to N_{i+1}. A sequence of these paths, $P_1, P_2, \ldots P_K$ connect the nodes of a chain. Each chain may be regarded as a single necklace. A combination of these chain paths will be called a global path.

The problem with the concept of segmentation is that it is borrowed from a programming language (PL/1), and, like Dijkstra, we would like to keep our (programming) structures independent of any programming language. Segmentation encourages arbitrary stringing-together of sub-functions or complexes, and thus may disaffect well-structuredness. However, under proper control, segmentation may lead to well-structured programs.

Another apparent weakness of segmentation structure is the absence of the concept of inequivalent functions (this can, however, be remedied). This notion will enrich our hierarchy structure and equip it completely for all uses in our development. It is the very basic principle for decomposing a complex. This concept is not widely mentioned by advocates of top-down techniques or other "structured" approaches. Inequivalence is a property of sets and functions, and may be introduced as shown on the following page.

Suppose that at a certain level of abstraction in the process of function decomposition, we obtain Δ_1 and Δ_2 as resulting composite structures. Then one of the following set-theoretic conditions holds:

$\Delta_1 \equiv \Delta_2$ The two structures are equivalent.

$\Delta_i \subset \Delta_j$ One structure is properly contained in the other.

$\Delta_1 \not\equiv \Delta_2$ The two structures are not equivalent.

It follows that any refinement of condition (1) will ignore one of the structures and decompose the other, since decomposing both would be duplicatory. Conversely, condition (3) will necessitate decomposing both other structures at the same level of abstraction. From condition (2) there arises more than one path at the same level of abstraction. We say that the two structures in condition (3) are inequivalent (complexes), but the two conditions in (1) are equivalent. Condition (2) is the most troublesome to refine, since at some level lower than Δ_1 and Δ_2 we may obtain substructures that are equivalent. In practice, this causes problems of inter-complex cohesion. Condition (2) proceeds with refinement of one structure, specifically the superset complex.

With these developments of inequivalent sets (functions, too, are sets), the process of hierarchical refinement proceeds as follows: At the highest level (level 0), we symbolize the problem complex as $\Delta^{(0)}$. This is the starting problem or the *root node* whose solution we seek to obtain. The refined components of $\Delta^{(0)}$ will occupy level 1. Therefore, at level 1 we have that $\Delta^{(0)}$ decomposes into some simpler subcomplexes as follows (Figure 5.3.2A):

$$\Delta^{(0)} \rightarrow \Delta_1^{(0)}, \Delta_2^{(0)}, \ldots, \Delta_k^{(0)}$$

i.e.,

$$\Delta^{(0)} \rightarrow \left\{ \Delta_j^{(0)} \mid j = 1, 2, \ldots, K \right\}$$

In popular language, $\Delta^{(0)}$ is called the father, and each $\Delta_j^{(0)}$, the son of $\Delta^{(0)}$. We'll use the following simple descriptive notation.

1. Subscript denotes the sequence number of refinement at current levels—counts the number of sons.

2. Superscript denotes the originating father node.

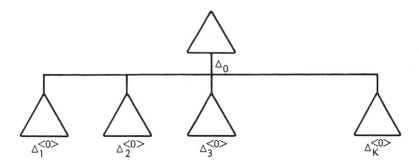

Figure 5.3.2A First-level refinement

3. The superscript of each complex will be a sequence of whole numbers. The last number in this sequence will be the subscript of the father complex.

From here on, the refinements increase in number depending on the complexity of the system. Each refinement is further refined at the next immediate lower level until the refinement terminates at some minimal functions. These minimal functions are critical for a well-structured system.

At level 2, each complex in Figure 5.3.2A may refine as in Figure 5.3.2B:

$$\Delta_1^{(0)} \rightarrow \Delta_1^{(0, 1)}, \Delta_2^{(0, 1)}, \ldots, \Delta_m^{(0, 1)}$$

$$\Delta_2^{(0)} \rightarrow \Delta_1^{(0, 2)}, \Delta_2^{(0, 2)}, \ldots, \Delta_n^{(0, 2)}$$

.

.

.

$$\Delta_k^{(0)} \rightarrow \Delta_1^{(0, k)}, \Delta_2^{(0, k)}, \ldots, \Delta_q^{(0, k)}$$

Each arrow indicates production of a set of inequivalent subcomplexes. The process continues until an irreducible complex results. Here the refinement terminates for that chain of subcomplexes. The algorithm repeats for each complex at level 1.

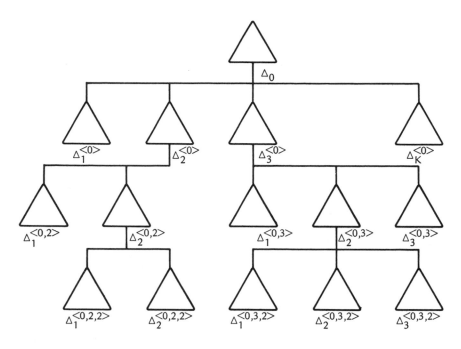

Figure 5.3.2B Second and third levels of refinement

In general (with a slight change of notation), each complex $\Delta_j^{(i)}$ at level j decomposes to level $j+1$, as

$$\Delta_1^{\langle n\rangle}, \Delta_2^{\langle n\rangle}, \ldots, \Delta_N^{\langle n\rangle}; \langle n\rangle = \langle\langle i\rangle, j\rangle$$

Finally, our initial complex, Δ, refines into the tree

$$T = \{\Delta_j^{\langle N\rangle} | j = 1, 2, \ldots, k; \langle N\rangle = \langle\langle N_{i-1}\rangle, J_{i-1}\rangle, i = 1, 2, \ldots, L\} \quad (5.3.2)$$

where L = number of levels, and K = number of sons.

The subscript of the father has been absorbed as the last element of the son's superscript. Clearly, $\langle n\rangle = \langle 0, n_1, n_2, \ldots, n_m\rangle$ is a *trace of the hierarchy* from the root node to the terminal complex of consideration. This is a very important development. *It suggests an alternative proof for the degree of structuredness.* In a well-structured module, every $\Delta_j^{\langle N\rangle}$ is unique.* The order of the sequence $\langle n \rangle = \langle 0, n_1, n_2, \ldots, n_m \rangle$, i.e., the number of nonzero integers, is called the dimension of the complex. This blend of structures yields what is called a tree structure, a derivative consequence of the hierarchical ordering of our basis structures. The dimension concept is a concrete reflection of the expanse and/or height of the "tree". It therefore follows that each complex occupies a topological point in the representative tree diagram. The notation (5.3.2) becomes the generalized notation of a tree. A two-dimensional representation of a tree structure is called a visual table of contents (VTOC), a software engineering equivalent of the traditional engineering blueprint.

Each minimal or irreducible function of a complex will be called a *monadic function*, or *monad* for short. Therefore, the final goal in the process of hierarchical refinement is to obtain all the ultimate monadic functions of a given complex.

What we have described so far pertains to functions. Programming also has the structuring of data. In it, there are two modes of structuring:

1. Dynamic structuring (as in Fortran, say) where disparate data are strung together at the instant of output; in general, this methodology defies hierarchical structuring.

2. Static structuring (as in Cobol, PL/I, etc.) where data items are described logically before actual storage.

EXAMPLE 5.3.2A: Dynamic Data Structuring—Fortran

```
    WRITE (5, 25) ISUM, RSUM, SIGMA, DELTA
 25 FORMAT (5X, I6, 4X, F9.4, 4X, F7.4, 4X, F12.4)
```

Each variable under the WRITE instruction may belong to a different buffer address or different file (area) within the module. That is, the variables are not required to be stored in a contiguous storage cell.

EXAMPLE 5.3.2B: Static Data Structure

```
01   CUSTOMER-INFO.
     03   CUSTOMER-NAME          PIC   x(30).
     03   CUSTOMER-ADDR.
          05   CUST-STRT-NO      PIC   9(5).
          05   CUST-CITY-ID      PIC   x(10).
```

---

*Some modules, naturally, may not yield to well-structuredness.

```
        05    CUST-STAT-ID          PIC   x(10).
        05    CUST-ZIP-CODE         PIC   9(5).
        05    CUST-THL-NMBR         PIC   9(9).
   03   CUSTOMER-EDUC.
        05    CUST-SCHL-GRAD.
              07    CUST-HIGH-SCHL  PIC   x(16).
              07    CUST-COLL-EDUC  PIC   x(16).
              07    CUST-OTHR-SCHL  PIC   x(16).
        05    CUST-DIPL-GRAD.
              07    CUST-HIGH-DIPL  PIC   x(5).
              07    CUST-COLL-DIPL  PIC   x(5).
              07    CUST-ETC-DIPLS  PIC   x(5).
```

In storage memory, this sequence of data names will be in contiguous location in this order, except perhaps for some linking (using pointers). The static mode encourages the grouping of data items according to their key significance. Care is needed at this design stage to ensure that the hierarchy here reflects that in-function hierarchy (whenever possible); the two should complement each other. For instance, in the design of reports, there is a bottom-up or top-down approach. The top-down approach designs the column headings before the individual data items of a record. Conversely, the bottom-up approach designs the columns of the individual data item first. The complementary approach designs the headings function before details functions such that the subfunctions correspond hierarchically.

One consequence as well as an advantage of good hierarchy structuring is the elimination of the THRU option of the Cobol PERFORM verb. It is a nuisance to good style because it contributes to excessive or unnecessary coding. Worse than that, it encourages the EXIT-IN-THE-MIDDLE (a camouflaged GOTO) habit of programming. It is a mere appendage to complier documentation and is not an active instruction.

Another consequence of good hierarchical structuring is the inadequacy of the top-down concept. There is another analogy of this approach in Dijkstra [56, p. 89] called nested virtual machines. Such an analogy, however, ignores the dimensional configuration of software, stretching out freely in N-dimensional space.

The introduction here of the Δ hierarchical structure as a basis will elicit some criticisms. However, the author's motivations for doing so are:

1. The hierarchical concept is about the single most important concept in structured programming and therefore deserves special emphasis as a basis element.

2. A complex of a system or architecture need not be decomposable into every element of the constituent basis, but into at least one of them.

3. The existence of chains of complexes with individual sequences of refinements best explains the hierarchical intrinsicness of a program.

4. Every piece of a complex belongs to exactly one of the several levels of the hierarchy. The hierarchy basis concept weaves these structures together.

5. The hierarchy concept comes close to the factoring concept of algebra where we sequentially break up complex problems into their most elementary groupings. It is an intrinsic attribute of a system.

In summary, we have introduced four structure elements as the bases for structured programming. These are the derivatives of any complex structure that is part of the definition of the conceptual system for solution. They are:

Σ — Sequence structure

Λ — Selection structure

Π — Repetition structure

Δ — Hierarchy structure

It must be cautioned that these elements, unlike those of algebra, are not linearly independent. Dimensions in computer systems don't share that property of algebraic systems because computer programs are generally composed of partial functions constructed on partial domains of data sets. Our use of them is purely for convenience of simplification.

Finally, having introduced the tree diagram, we can further observe that IF-tree diagrams are comprised of subtrees which fit in naturally with the general tree, in addition to indicating the richness or complexity of our representation. Most programmers recognize the difficulty with nested-IFs—the deeper the nesting, the more complicated the programming and debugging. Having shown that this structure is a binary tree and having also shown that the generalized SELECT structure has the general tree diagram, it is clear that the use of the tree diagram to define a system's complexity is necessary and sufficient.

The single-entry/single-exit feature of Mills is a necessary concluding feature of Dijkstra's hierarchical theory. This feature enables us to know where to begin and where to end processing a complex. From another perspective, the sequence of single-entry/single-exit can be seen as represented by the trace of hierarchy, $\langle n \rangle$. Since the author considers it a sine qua non of the hierarchical notion, the two together are treated as a single basis element. This further strengthens the argument of stripping off this feature from conventional "structured" flowcharting in order to keep the syntax separate from the semantics of structured programming.

In summary, with this basis set defined, we may now answer the question, what is structured programming?

A *structured module* is one that is composed of the set of four basis principles.

Structured module (program) $\triangleq \langle \Delta, \Sigma, \Lambda, \Pi \rangle$ where the symbol $\langle \rangle$ indicates generation, as is used in algebra.

A *well-structured module* is a structured module in which the degree of structuredness = 1. (This is the ideal state of structuredness).

Structured programming is the rigorous application of these bases principles in the development and maintenance of software.

The remarkable consequence of these two definitions allows a range of structuring modules, particularly as regards the degree of hierarchical structuring.

5.3.3 THREE FUNDAMENTAL ALGORITHMS FOR STRUCTURED PROGRAMMING

The algorithms discussed here will reflect our developments of the fundamental and the hierarchical structures. In the end, these algorithms do more than vindicate our earlier speculation that there does exist a (scientific) methodology for analyzing, designing and realizing software systems; they also justify the term software engineering.

5.3.3.1 Algorithm for Fundamental Control Structuring

1. Determine the nature of the problem.

2. Determine and classify any available concepts about solutions.

3. Determine a feasible solution from (2).

4. Determine a suitable programming language.

5. Determine and classify the structures of propositions and predicates:

    a) Simple propositions—lead to sequences

    b) Complex propositions—composed of cuts. Logical connectives—chain of sequences; conditional connectives—select structures; quantifiers (universal or existential)—loop structures, etc.

EXAMPLE 5.3.3.1A: Given the following data:

3, 6, 7, 4, 5, 6, 9, 0, 2, 4, 4, 6, 7

compute the mean, median, mode and the first and third quartiles.

    Step 1. This is a problem in statistics—computation of static, mean, median, etc. Each of these has a formula, not reproduced here. The interested reader should consult elementary texts in statistics.

    Step 2. The first computation will be the mean, since some of the other statistics depend on it; care should also be taken to obtain a count, N, of the number of data elements. Matrix storage of the input data will be helpful.

    Step 3. A suitable programming language is Fortran or Algol or BASIC. Don't ask for Cobol.

    Step 4. The simple propositions are: COMPUTE STATISTIC, S; IMPLICATIONS are TESTS of conclusion of a given computation; this is signaled by the COUNT N as well as the repeats of ADDITIONS.

Thus the fundamental structures are:

1. Sequence—compute statistic, S

2. Select—*if* count greater than N, then do desired function or else continue.

3. Looping—do activity until count = N.

5.3.3.2 *Algorithm for Hierarchical Function Structuring*

The algorithm for function structuring or the refinement of complexes is as follows:

    Step 1. Identify all major functions or activities at the zero or initial level.

Step 2. Classify these functions into equivalence classes. These form the complete chain of complexes for the module (system) concerned.*

Step 3 For each class in Step 2, define its equivalence subclasses, and repeat this until all the monads of this chain have been obtained.

(One may recognize the recursive nature of hierarchical structure.)

EXAMPLE 5.3.3.2A: This example is excerpted from a published example. It is typical of many "structured" Cobol programs.

```
MAIN-LINE.
    OPEN   INPUT    DATAFILE.
    OPEN   OUTPUT  PRNTFILE.
    MOVE  'N' To EOF.
    PERFORM READ-FILE   UNTIL   END-OF-FILE.
    DIVIDE  TOTAL  BY  N  GIVING  MEAN.
    MOVE  MEAN   TO  PRNT-MEAN.
    WRITE  PRNTREC
            AFTER ADVANCING 2 LINES.
    CLOSE DATAFILE   PRNTFILE.
    STOP  RUN.
```

To give structure to this program, we begin by resolving the problem of equivalence of functions, and then proceed as follows: The initial level in equivalent functions are: (1) Open; (2) Read; (3) Write; and (4) Close. MOVE and DIVIDE are strictly subfunctions and must not appear on this level, that is, the first level of this module has four inequivalent functions, and hence there are four chains:

```
PERFORM   OPEN-FILES.
PERFORM   READ-FILES.
PERFORM   WRITE-RECORD.
PERFORM   CLOSE-FILES.
```

This allows us to add or modify subfunctions within any chain.

-----

*Inequivalence is the fundamental property that determines independence of modules/systems.

EXAMPLE 5.3.3.2B

Assume that a program is to be designed to compute mean, M, and standard deviation for a class. We wish also to assign letter grades to individual student scores. The input to this program will be student identification grade cards: id# and score. The rules for assigning letter grades are as follows:

If S is greater than	But no more than	Letter grade
$M + \frac{3}{2}\,a$	100	A
$M + \frac{1}{2}\,a$	$M + \frac{3}{2}\,a$	B
$M - \frac{1}{2}\,a$	$M + \frac{1}{2}\,a$	C
$M - \frac{3}{2}\,a$	$M - \frac{1}{2}\,a$	D
0	$M - \frac{3}{2}\,a$	F

Solution

If the information above is complete, the inequivalent functions are:

1. Read grade cards
2. Calculate mean (score)
3. Calculate standard deviation
4. Assign grade.

(We have for now ignored opening and closing of files.) These then are the nodes of chains of complexes at level 01. Of special interest is the 4th chain. We already have the necessary information to obtain further refinements at level 2. It is a generalized SELECT function among A, B, C, D, E. Each has a range of values between the boundries of greater than and not more than, as shown on the table.

5.3.3.3 Algorithm for Hierarchical Data Structuring

Step 1. Determine all the aggregates of data in the given system (module).

Step 2. Apply the principle of equivalence to partition (1) into classes; call each class a record at this level.

Step 3. For each record, apply the principles of equivalence to discover its subclasses.

Step 4. Determine a grouping device for related or chains of equivalance subclasses.

Step 5. For each group in (4), repeat step 3 up to irreducibility; each irreducible element will be called an *atom*.

Step 6. For each atom, determine its data attributes—size, type, usage, etc.

EXAMPLE 5.3.3.3A: Hierarchical Data Structure

Suppose in a system involving an oil well, information is to be designed according to segments corresponding to land ownership, production frequency, government regulations, windfall profit tax, shipping regulations.

Step 1. The aggregates of data are ennumerated here under assumptions of land ownership, etc. However, in practical applications, these aggregates will be determined by the analysts.

Step 2. The equivalence classes are predetermined explicitly here. However, land ownership may be split into inequivalence classes.

Step 3. Using the aggregates determined by the analysts, each record may be defined.

Step 4. An instance of grouping may utilize government regulation records: state attributes, claimer attributes, age of well attributes, tax attributes, etc.

Step 5. For each element identified in Step 4, we apply the principle of equivalence repeatedly until we obtain all its atoms.

Step 6. Assigning data attributes to these atoms may call for consultation with a systems or business analyst.

We can thus talk meaningfully about the degree of complexity of data (structure). It too is representable as a tree diagram.

5.4 RAMIFICATIONS OF THE STRUCTURED PRINCIPLES

———————◆———————

No revolution takes place without ramifications felt in its environment and even beyond. Since the applications of structured programming have been popularly acclaimed in the computing industry, its practices have diffused into other areas of the programming profession. The philosophy has embraced such activities as analysis, specification, design, and documentation; it is therefore justifiable that the philosophy of structured programming admits these phases of the industry as complimentary areas of applications. If a system is poorly analyzed initially and/or is ill-designed, then its structure programming will be grossly disaffected. The hierarchical basis principle must be initiated at the inauguration of the system's life cycle.

The same principles that apply locally to programming also apply globally to analysis and design, though at much varying degrees of detail. This section introduces the following applications of the structured revolution: structured analysis, structured design, structured documentation, and structured specification.

The word programming in structured programming is more than just coding and testing. Structured programming has become a philosophy that encompasses nearly every human activity of the computing process in the information industry.

5.4.1 STRUCTURED ANALYSIS

The use of sketches and similar illustrations is the rational beginning of the analysis phase of a system. This is the primitive phase of the system's life where concrete ideas are derived from conceptual systems thinking using the methods of abstraction. It is from these sketches and illustrations that data flow diagrams (Figure 5.4.1B) are derived. However, the use of data flow diagrams alone as structured analysis is not consistent with the development of hierarchical structures. Flow diagrams are actually tools derived from data-driven methodologies. Many of them with their several characteristic crossing lines resemble the familiar spaghetti structures. Data flow diagrams are "functionlike" (Figure 5.4.1A) when considered individually, but their unstructured concatenation may lead to confusion and disorder.

Figure 5.4.1A   Function: A function has three components: DOMAIN—From-where; DESCRIPTION—What; RANGE—To-where.

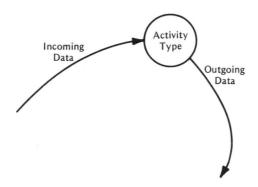

Figure 5.4.1B Data Flow Diagram: A data flow diagram has three components: INCOMING DATA; ACTIVITY TYPE; OUT-
GOING DATA.

Although data flow diagrams have become widespread, to be consistent with the principles developed here, it is necessary to replace them
with well-defined *function diagrams*. Function diagrams have the additional attributes of better numbering (or structure labelling) and control.

The average systems analyst is familiar with data flow diagrams. This
is probably so because the user is concerned with his or her data, data
types and files. Consequently, at the interview stage with the user, the
data flow diagram is the most concrete tool the business analyst can use.
But, behind the user contact, the competent systems analyst must begin
to transform this communication into structured analysis. Traditional
data flow diagrams, like flowcharts, have the unequivocal capability to
do more harm than good. They must therefore be refined and drawn in
such a way as not to resemble spaghetti diagrams. A good idea can be
drowned in a bad diagram; clarity of thought must be reflected in simplicity of diagrams. The analyst must rewrite his or her communications
to simulate computer capabilities for computations. This is where the
entity-relationship concept is most useful. As a powerful analytical tool,
it can be used to disentangle a mix of ideas and thus violate the individual functions.

This involves refinement of the data flow diagrams using notes,
sketches, memos and other communications. In this process, the center
of attention should be the well-formed functions, where the various functions are defined as well as their corresponding data domains and ranges.

When not so defined, the analyst must compensate by a suitable transformation for the benefit of the designer.

Data flow diagrams are really camouflaged representations of functions. For instance, an UPDATE function is associated with some input and output data. Thus the definition or extraction of the UPDATE will identify all its operating data. At this stage, it is then possible (for the design analyst) to decompose the UPDATE function down to its monadic functions.

With the conclusion of function analysis, the next analysis responsibility is to compile the data dictionary. Some authors suggest this should be done at the completion of data flow diagrams. However, it may be useful to have them done as suggested here, so as to include in the data attributes the purposes for which the data (items) are to be used. In defining the elements of the data dictionary, consideration should be given to the "principles" of structured coding.

In addition to the data dictionary is the definition of data storage. This is the responsibility of the systems analyst who knows the details of internal computer storage. The necessary tool here is not functions per se, but the hierarchical basis for data structures, which compliments function hierarchy. The association of data with functions clearly justifies preparation of the data dictionary after function analysis, and not after data flow. Where structured data bases are in use, the systems analyst may work with the data base staff to define the schemas and subschemas.

At this stage, it is clear that the functions identified so far are representative of the functional procedures and purposes of the organization as reflected in the requirements defintions of the system. Their various impacts on the system as a whole is the central concern of the design analyst and must be conveyed explicitly by the business analyst.

The final responsibility of the systems analyst is producing the systems specification. It is assumed that by this phase of the system's life cycle documentation (including data dictionary) is almost completed.

5.4.2 STRUCTURED DESIGN

The design phase should begin with sketches using the analyst's systems specification to realize the set of functions (implicit or explicit) to be processed. Once these functions are fully realized, the emphasis shifts to identifying their intricate relationships. At this stage there is not much emphasis on data, since the data are assumed to be part and parcel of the functions, i.e., attention is on functions. It is when we shift attention to the individual functions that the totality of data (belonging to that individual function) comes into consideration. Furthermore, it is advisable that the first sketches of the design be made independent of any programming language. The programming language should become prominent when it is time to write the design specification, for then the notations of that language will come in naturally. Structured programming must remain independent of programming languages. This means that the design specification just about concludes the design phase.

Structured design follows the same principles developed earlier in the chapter. Naturally, the first set of functions realized form the first level of abstraction. They're sketched as in Figure 5.3.2A. This is the first step toward construction of a tree diagram for the module. It follows naturally from our processes of realization; for each such function, we examine its data elements in some detail.

We can sketch the function's inputs and outputs as in Figure 5.3.2B where the box represents the function. Now we are ready for the second level of abstraction. We know that a function may be a composition of other functions (actually subcomponents). We identify the totality of these (sub) functions for the particular "major" function (node) under consideration. Then the process of sketching its branches may now be completed as the second level of abstraction. It is, of course, assumed all the data sets for each subfunction are fully realized.

At the third and subsequent levels, the above processes are repeated until a given component function can be refined no further. This is one *monad*, and any number of them may have been derived from the starting chain (of complexes).

The next iteration takes us back to the first level where we repeat the above processes from level two for each function at the first level of abstraction. This completes the graphical design—yielding the tree diagram of our module—and sets the stage for design specification.

We remark here that if a complex decomposes, it must decompose into at least *two* inequivalent subcomplexes and up to monads. Thus Figure 5.4.2C is an invalid refinement.

Next, it is necessary to show the trace of hierarchy or single-entry/single-exit flow patterns of the tree diagram. Figure 5.4.2B shows these flow structures by means of directional arrows. Diagrammatic convenience seems to use the same path, but this is not the case in practice, where we "see" the entry at the "physical" top of the box, and the exit at the "physical" bottom. However, this diagrammatic convenience enables us to tie loose strings together—a matter of accomplishing a GO-

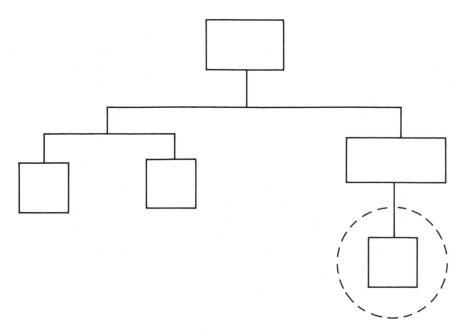

Figure 5.4.2A AN INVALID REFINEMENT. The parent of the circled
node is ill-refined.

BACK (as in Cobol) to exit from where we started, thus completing a
circuit. In the end, we know where to go and where to return. This is
structured thinking.

Finally, having shown that the nested-IF structure is actually a binary
tree, we notice then that our basis structures can also be represented on
the tree diagram. Both the sequence and iterative structures appear as
"function boxes," while the hierarchical structure is just "architecturally"
visible throughout the length and breadth of the diagram. We now call
this tree diagram the *software structure diagram* to distinguish it from
VTOC with its two-dimensional limitation. The software structure diagram now becomes the ultimate design blueprint of software engineering.

In summary, let's point out one functional difference between structured analysis and structured design. While the latter concentrates on
the technical relationships of global and local data flow within and without of the system, the former emphasizes and delineates the global flow
of systems logic as well as choices about physical devices for data storage.

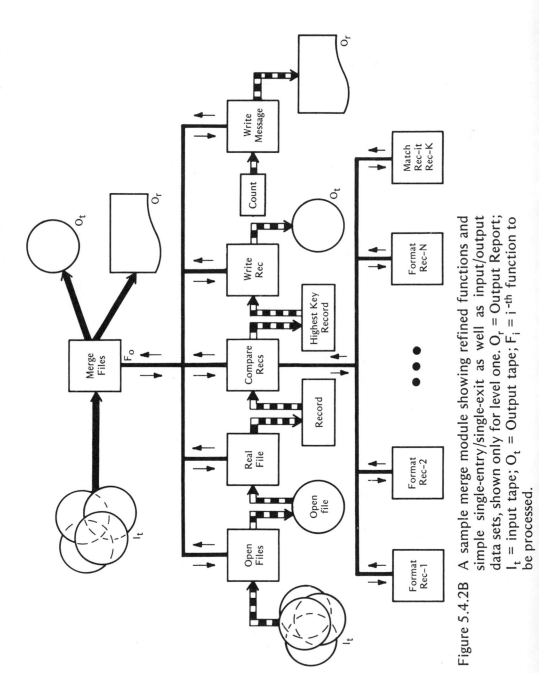

Figure 5.4.2B A sample merge module showing refined functions and simple single-entry/single-exit as well as input/output data sets, shown only for level one. O_r = Output Report; I_t = input tape; O_t = Output tape; F_i = i-th function to be processed.

5.4.3 STRUCTURED DOCUMENTATION

The documentation discussed here will focus only on the technical level, utilizing the software structure diagram. This is the only one of the four levels of documentation that ties in closest with the function concept, and, is the most important aspect of the four levels for maintenance purposes.

Documentation can be fun if the design progresses along the principles discussed earlier and if both the design and analysis have utilized the structured concept. In that case, we will be documenting the "modular" functions. There is no exaggerated claim that documentation of Figure 5.4.2B which explicitly designs the functions, showing the input and output data for each, leads to a satisfying goal.

The use of IBM's documentation package (HIPO form and HIPO template) comes in here as an important aid to documentation. The strategy of representing all the functions refined from a given complex on pages helps visual recognition. Then each function is documented on a separate page. The HIPO form provides for notes or remarks. Very important in this technique is the numbering system, or the structure labels. Some programming languages allow these structure labels to be coded inside the source module itself. This makes the module's contents consistent with its external documentation. Structure labelling, in the end, is a form of structured documentation.

The philosophy of "document as you design" is consistent with structured programming. However, in practice, the professional must match practice with realism, or expediency at least. The theory behind using the principles of structured programming to document is itself self-explanatory and sufficient. It is hoped that "document as you design" is valid only for those aspects (functions, etc.) that have been concluded. Documentation and design should as much as possible compliment each other.

Above all, the programmer should avoid over- and underdocumentation. The function approach minimizes both of these. Be concise—simply show data sources and destinations, together with brief descriptions of what the function does. Complicated algorithms may be described elsewhere, outside the documentation pages. Finally, minimize excessive use of paperwork. Even museums do run out of storage rooms.

For efficient documentation, there are documentation symbols for software design. To be consistent with the notational symbols of the Böhm-Jacopini basis elements (Σ for sequence; Λ for selection and Π for iteration) we write each symbol at the top left corner of its corresponding function net. Thus the new design documentation symbols for the fundamental structures reduce to Figure 5.4.3A. Two major consequences of the new notation are explained below.

5.4.3.1 Single-Entry/Single-Exit Pattern

The device chosen to represent single-entry/single-exit may be controversial, but the concept is logical. The statements of a sequence (rectangle) are arranged sequentially from top to bottom. Some of these rectangles are at the lowest level, and the (algebraic) law of closure forbids exiting at their bottoms; thus we must *return* to the entrance door to exit—completing a circuit. In essence, we enter at the "door," navigate along the perimeter of the rectangle and return. In addition, this device enables us to enter at the top of the next sequence.

However, a word of *caution*. At the coding phase, a programmer might violate this single-entry/single-exit principle by offending the axiom of super-to-subordinate communication being a relationship of a father-to-a-son (see axiom 5 in 3.1).

5.4.3.2 New Structure Label for Three Nodes

For consistency, our new notation has effected a necessary change in documentation of the tree nodes. The documentation introduced by IBM in the form $N_1.N_2.N_3, \ldots, N_2.0$ is no longer effective. Our structure label, the trace of hierarchy, is a sequence of integers and this property (of sequence) must be reflected in the documentation. In mathematics, the elements of a sequence $\langle N \rangle$ are separated by commas. The IBM method documented the nodes at level 1 as 1.0; 2.0; 3.0; 4.0; etc. The new method becomes 0,1; 0,2; 0,3; 0,4; etc. Then at the second level the derivative nodes are, respectively:

$$0,1 \rightarrow 0,1,1; \quad 0,1,2; \quad 0,1,3; \quad \text{etc.}$$
$$0,2 \rightarrow 0,2,1; \quad 0,2,2; \quad 0,2,3; \quad \text{etc.}$$
$$0,3 \rightarrow 0,3,1; \quad 0,3,2; \quad 0,3,3; \quad \text{etc.}$$
$$0,4 \rightarrow 0,4,1; \quad 0,4,2; \quad 0,4,3; \quad \text{etc.}$$

Each node reflects its originating root, 0, as the first element of its sequence.

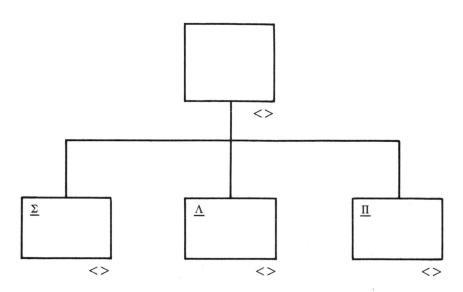

Figure 5.4.3A Design documentation symbols. Using *bases* structure symbols to annotate function structures: for SEQUENCE; for SELECTION; for ITERATION.

5.4.4 STRUCTURED SPECIFICATION

In the late '70s, there was a move to invent a language called specification language for writing specifications [8] or problem statement language (PSL) [47] or semicode. These devices are oriented toward model building—a (mathematical) description of the set contents of a system. This is evidence of dissatisfaction with the existing languages for documentation. However, the author joins the critics who question the need for the proliferation of programming languages. It must be admitted that some of these programming languages (e.g., Cobol) lack suitable notational symbols for specification. However, what is needed is an evolutionary refinement of the existing ones. It is therefore necessary to keep the academic interest going for purposes of research. The present status of problem documentation languages (PDL) is hardly sufficient for practical purposes, except perhaps to contribute to overdocumentation.

Structured specification is strongly tied to function thinking, since specification is a mechanism for defining the activities to be accomplished (inside the computer). Since a function is defined on data sets, each function relates some *input* data to its corresponding *output* data. The specification may therefore use one of the conventional notations of functions

$$\text{INPUT} \xrightarrow{F} \text{OUTPUT}; \quad \text{or}$$
$$F : \text{INPUT} \rightarrow \text{OUTPUT}; \quad \text{or}$$
$$F = (\text{INPUT}, \text{OUTPUT})$$

or any other suitable notation. Every good specification must be precise and unambiguous; this is why the use of functions is encouraged. A function has a list of arguments (inputs) and a set of operators (definitions of computations). It may be necessary to use set notation to show the members participating in a function.

Of the four basis set of principles, structured specification utilizes the Böhm-Jacopini basis set more than the Dijkstra-Mills basis. This is because of the emphasis on the selective decisions to be made—a theoretical expression of writing specifications (both analysis and design) in English-like notations, using data names as defined in the data dictionary. It is the combined use of these data dictionary names and the structure basis elements that has given rise to the term structured specification.

Structured specification is used for both the analysis and the design phases (sometimes called pseudo-coding) of a system. But the content of each determines for what phase and purpose it is written. Excluded here are narratives and such memorandalike descriptions that are erroneously called specifications; at least, they are not structured. An analysis specification, for instance, uses more businesslike or conversational forms, while a design specification uses more computer operation terminology. Scientific systems use heavy notational and set symbols.

However, serious (structured) specification is little practiced in many commercial environments. There, specification sometimes merely means a documented outline of a project, or is even interpreted by some as equivalent to narrative. This is erroneous. Furthermore, in some smaller to medium-sized installations, design and analysis (of modules) seem to progress simultaneously, if possible by the same person. It is obvious that bugs may be introduced in such a module, except if the professional is especially careful. In these cases, the line of demarcation between analysis, design and realization is blurred, and some details may be omitted. For example, specifications may not be produced, and when they do, they may be scanty in details with perhaps no structured principles—even in this era of structured programming.

EXAMPLE 5.4.4A: Semi-Code/Systems Analysis

```
OPEN FILES.
BEGIN INPUT DATA FROM FILE-J.
IF RECORD TYPE OK
   BEGIN VALIDATION
ELSE
   PRINT MISCELLANEOUS MESSAGE.
IF LAST FILE RECORD ACCESSED
   END INPUT DATA.
IF ALL FILES READ
   DO MERGE FILES
ELSE
   BEGIN INPUT DATA OF NEXT FILE.
BEGIN SORT FILES
AT END CLOSE FILES.
STOP.
```

EXAMPLE 5.4.4B: Semi-Code/Systems Design

```
ESP-DESIGN-SPEC.
     DO   OPEN FILES.
     DO   PROCESS-INPUTS-UNTIL   ALL-END
     DO   SORT-RELEASES.
```

```
DO   CLOSE FILES
STOP RUN.
OPEN FILES.
    OPEN INPUT F1, F2, . . . , FK
        OUTPUT FM, FN, . . . , FT.

PROCESS-INPUTS.
    DO   READ-FILE.
    DO   VALIDATE-REC-TYPE.
    DO   FORMAT-RELEASE.

SORT-RELEASES.
CLOSE-FILES.
READ-FILE
VALIDATE-REC-TYPE
FORMAT-RELEASE.
```

A systems design specification is supposed to be more detailed and precise.

EXERCISES

5.1   Name and describe common functions in software environments. Show that each may be "specified" analogously to conventional mathematical notations. Tabulate the advantages and disadvantages of function-oriented thinking in software systems development.

5.2   Show the differences between recursion and iteration. What conditions necessitate these applications?

5.3 (a) Show that the structures DOWHILE and DOUNTIL can be shown to be "equivalent" as to elements.

   (b) Describe an alternative proof that the three fundamental structures are complete for the decomposition of any program complex.

   (c) Prove that $D_s = 1 \Leftrightarrow$ every $\Delta_j^{\langle N \rangle} \epsilon T$ is unique and

   $$\exists \left\{ \delta_k^{\langle n \rangle} \mid k > 1 \text{ and } \delta_k^{\langle n \rangle} \epsilon \Delta_j^{\langle N \rangle} \right\}$$

   such that $\delta_k^{\langle n \rangle}$ is irreducible. We call this the *fundamental theorem of software design*.

(d) Show that if S is structured, then

$$\exists\, T = \left\{ \Delta_j^{(N)} \mid j > 1 \right\} \ni S \cong T.$$

5.4 (a) Tabulate the steps for a system analysis.

(b) Show that structuredness is incomplete without the single-entry/single-exit feature. Do the same for the independence property.

(c) Tabulate the uses and advantages of documentation.

(d) Demonstrate by examples that some programming languages allow more powerful structured specification than others.

(e) Show that the structure label $(N_1, N_2 \ldots, N_n)$ satisifies the polynomial

$$t_n(x) = \sum_{i=0}^{n} N_i x^{i-1}$$

up to degree n. Show further that the set of polynomials

$$T_L(X) = \left\{ \langle t_n(x) \rangle \mid n = 0, 1, 2, \ldots, L \right\}$$

generates an arbitrary tree; $T_L(X)$ will be called the tree polynomial.

PART TWO:

METRICS of SOFTWARE ENGINEERING

The great gulf between Art and Science is the precision of measure.

Art describes and accumulates; Science defines and measures.

Chapter 6

A NUMERICAL MEASURE for PROGRAMMER PRODUCTIVITY

Counting has long been with us. Nearly everyone does it, but does everyone really know how to do it? Though the whole thrust of mathematics is concerned with counting, the many aspects of counting are so challenging and complicated that there is more ignorance than knowledge about it.

Comparing is also a form of counting because it ultimately leads to some form of counting. But this is not always apparent. Without the faculty of counting, and without the ingenuity of comparing, man would not have invented numbers. Numbers are the final invocations of mathematics. They are an artificiality, the aftermath of the thought processes. They are an abstraction [30, p. xiii].

6.1 THE NEED FOR STRUCTURE ON SOFTWARE SYSTEMS

◆

Structure is the observable or conceivable arrangement of an object. The higher the expression of structure, the more refined the object possessing that structure.

However, measure is an appraisal of structure. It therefore follows that the initial study of an object for determining measure is the *structure* of that object. In the absence of a well-defined structure, there is nothing to measure. For geometrical objects, for example, the line surfaces are clearly defined; for algebraic surfaces, the equations of the surfaces are used.

117

But how does one go about measuring intangible complexes? Even though a measure for information (a "product" of software) is known, the measure for those products or means that produce information has lagged behind. The simple answer is that software engineering has traditionally lacked structure. There have been no well-developed theories about the structure of software. The explosion of applications utilizing fast computing machines left no time for research into methodologies of applications. The computer revolution descended with an explosion of concepts, languages, problems and solutions.With them came an avalanche of confusions and frustrations.

Another major setback in the development of feasible measure is the slow recognition by commercial programming professionals of the concept and power of functional thinking. (Chapter 5 dealt briefly with this.) It is ironical (?) that the Böhm-Jacopini developments utilized some mathematical (functional) reasoning to give us what is here called the canonical (accepted) basis for structured programming. According to Dijkstra, mathematical reasoning enables us to grasp complex structures with our limited mental capacity [16].*

This reluctance to use the functional approach had its roots in the hazy polemics about what science is or what art is. But the question is, Can something be both a science and an art? Some people considered programming an art. Recently, modern developments are bringing it to a science. It was really never an art. We must recognize the strong and undeniable influence that language and environments impose on our thoughts and habits. They tend to limit our ability to abstract and formulate ideas. However, this traditional proclivity is not inescapable. If there is any distinction between art and science, it is this: the techniques of science follow a set of well-defined principles. It is the absence of these principles that tended to make programming an art. The techniques of an art have an inherent power of the occult. He who would learn an art must commune in some way with the initiates; but he who would learn a science must commune with the principles. In the absence of well-defined principles, a study or theory or philosophy of anything is simply an art. The immediate consequence of introducing well-defined principles is the introduction of a precise measurement scheme. According to Clark Maxwell, to measure is to know.

Another diverting influence from the search for structure is the search for more programming languages. The influence of language on the programming community has sometimes been traumatic. Many people are wondering if there are not too many programming languages already [32]. Dijkstra's reflections following a symposium about higher level programming languages [17], on the cumbersome size of the PL/1 language because of its extensive requirements for more features, best describes this concern.

Although certain programming languages have capabilities lacking in others, of most concern in the theory of structured programming is the intellectual management of complexity. When this complexity is further complicated by a progamming language, our anticipated success is greatly diminished. Fortunately, the developments of the structured revolution have been independent (except for minor infractions) of programming languages. These languages sometimes impose such harsh constraints on us—both on the control of physical outputs and construction of software components—that the very problem we are attempting to solve becomes "dedicated" to a particular language definition. This adds to complexity.

EXAMPLE 6.1A: Fortran IV

```
        WRITE (6,36) ALPHA, DELTA, SIGMA

36 FORMAT (3(4X,F12.8))
```

This language does not require the programmer to rigorously specify I/O files; the number 6 is a "sufficient" designation.

EXAMPLE 6.1B: Cobol

```
ENVIRONMENT DIVISION.
INPUT-OUTPUT SECTION.

FILE-CONTROL.
    SELECT STRCFILE ASSIGN TO UT-S-STRCFILE.

DATA DIVISION.

FILE-SECTION.
FD  STRCFILE

    LABEL RECORDS ARE STANDARD
    RECORD CONTAINS 100 CHARACTERS
    BLOCK CONTAINS 10 RECORDS
    DATA NAME IS STRCFILE-REC.
01   STRCFILE-REC                              PIC X(100).
PROCEDURE DIVISION.

WRITE-EXAMPLE-FILE.
    WRITE  STRCFILE-REC    (AFTER ADVANCING 2 LINES).
```

Time was needed for experts and pioneers to tackle the problems, and the success made so far has been encouraging. The developments of structured programming theories will lead the way to the future solution of these perplexities.

A well structured module is incomplete without the definition of a degree of software *structuredness* developed in 7.2. Furthermore, time has come to define the familiar headache of software engineering—the degree of complexity. One such definition is postulated and presented in 7.1.

Finally, as feedback, the formulae developed in chapter 7 will reveal some of the latent characters of software systems which have so far been manifested as bugs and degraded productivity. In particular, what is called the Coefficient of System Behaviour (with the participating factors presented in chapter 7) vindicates this claim. This is the great value of measure; when we measure, we know.

However, I did not set out to find or identify these characteristic behaviors. Arriving at them has been a consequence of the law of serendipity. The emphases throughout these developments have revolved around the tree structure of 5.3.2.

Any science must have its own method of investigation, and software engineering hitherto had no other such method than the Software Structure Diagram.

6.2 SOFTWARE ENGINEERING
BECOMES A SCIENCE

The criticism that the evolution of software has lagged behind that of hardware emphasizes the absence of suitable measures defined for software. This criticism has also been carried further so as to discredit software as a science. Many branches of science and engineering have well-defined and elaborate metrics for calculating their operating factors, especially cost and productivity. Just as computing hardware has revolutionized industries, our software measures for productivity will also be found useful in more and more industries. Perhaps the impact of software engineering on hardware engineering through structured programming may lead to a broader revolution in engineering—structured engineering. The simplex algorithm of section 5 will have one such impact.

The emphasis in this section is on those simple measurable attributes of software. Some of these factors are obtainable by counting, e.g., levels or dimensions, number of requirements definitions, SELECT structures. More than that, the equations are automatically computable. Hardware and management metrics are excluded for the sole purpose of not adding to the existing confusion in software metrics.

Traditionally, cost and productivity have been painful concerns in computing environments. The factors entering such calculations have been based on personal hunches and heuristics, not scientific or feasibly reliable techniques. A user is charged a specific fee for certain programs (development, coding, testing, etc.) based on fixed units arbitrarily "manufactured" by some department. The consulting industry, too, bills services based on nontechnically documentable figures. Some consultants charge $25 an hour, others $50 an hour. But can they provide a rational scale of measure of productivity to justify the assessed cost?

But most crucial in this argument is the programming manager's approach to assessing his or her daily workload. How does the manager evaluate his or her staff? How are such evaluations ensured to be consistent and unbiased? Is there any documentation to back up evaluation factors? Presently, these questions are unsatisfied. Whatever evaluation measures exist are drawn up by personnel, and usually these evaluation criteria are no different from those used for clerical or labor staff, except perhaps for some sophisticated terminology.

Besides these estimates of cost and evaluation, better knowledge of programmer productivity may help in organizing the computing center: planning human and material resources, scheduling development, coding, production runs.

A traditional recommendation is to use lines of code as a measure. The author is satisfied that this method is at best adequately inadequate. For unstructured programs or software, this may be a crude substitute, but no measure at all is actually better. The only way, perhaps, to make this method meaningful is to count all the non-blank characters per line; however, this is very time consuming. Another reason for rejecting this measure is that no two programmers code alike. What one programmer may code in 400 lines, another may do in 325 or even 200 lines. Who of these two is more productive? Surely, the experienced programmer will do it in less time (possibly) and in less number of lines (obviously).

Another variation of the lines of code measure is to count operators and operands, but it too suffers from programmer incompetence.

Consequently, the effort here has been to lay down some viable factors possessing attributes for obtaining concrete metrics on software. The strategy has been to avoid rigor and resort to elementary arithmetic, such as direct variation and inverse variation of a factor. If a factor under study increases, what effect does it produce on its target of study and vice versa?

This notion is used to develop the significant factors of productivity: degree of complexity, degree of structuredness, degree of requirements definition, etc. Some of these factors have never received any consideration in the literature, but they are critical in understanding software. It seemed rational and logical to develop these measure factors on the attributes of a well-known abstraction, the software structure. There is no hypothesis, even in the author's theory of reliability, that invokes any principles outside the environment of software engineering.

Finally, there are in practice numerous factors that are mentioned in the theory of programmer productivity which the author has not found relevant for inclusion in these formulae. Some of these are really subfactors of the ones singled out here as significant, while others are inconsequential to these measures. The former includes such topics as influence of user contact (absorbed in the degree of requirements definition), structured techniques (absorbed in the degree of structuredness), and availability of resources (absorbed by turnaround time). The latter include time spent at meetings and conferences, choice of programming language or computing hardware, etc. The third class of factors are the psychological variable "invariants" of mood, health, weather influences, preference for work, etc. They are not considered in this development of the software metric system.

6.3 GENERAL FORMULA FOR PROGRAMMER PRODUCTIVITY

───────────◆───────────

An effective measure of (programmer) productivity must be expressible in terms of work done. In mechanics, where the concept and practice of measure are well defined, work is a measure of force and distance moved; in electricity, power is a measure of voltage (which is a form of force) and resistance or opposition to doing work. It follows therefore that a meaningful measure of productivity must analogously involve time duration, some notion of function (which replaces the moving force of mechanics) and any other weighting factors. To be specific, it is justifiable that for our function we consider the aggregate definitions of activities comprising our inputs/outputs. For us, these "functions" are explicitly defined in a well-structured software; in particular, the hierarchical refinement of functions as expressed in a software structure diagram will be a valuable fundamental asset.

There are two types of measures for programmer productivity. The one suggested by Yourdon [52] is a cost measure. Unfortunately, its usefulness is very limited since it cannot be used until the conclusion of the project so as to determine the respective costs for analysis, design, coding, testing, debugging and reliability. Yourdon calculates the "effectiveness" of the program and, of course, the programmer by summing these factors weighted by appropriate (but aribitrary?) coefficients.

The other method introduced here is a *progressive evaluation technique* (PET). Unlike most traditional methods, it requires no paper work, no logs, no graphs, scales or charts, and is therefore no additional clerical burden on the user. In practice, some of these factors may be effective coefficients for the Yourdon cost equation. Our strategy will utilize resources at our command—the structured software before us. It is assumed the program is at least (well) structured. Furthermore, we shall normalize these coefficients to lie between 0 and 1; the idea is to obtain not too large a number, but a convenient one. However, the user may multiply by 100 or as desired. The following factors of software will be used in the calculation:

1. Functions—monadic or minimally refined functions

2. Time—duration of concluding some "task" or project or assignment

3. Weighting factors: (a) degree of complexity, (b) degree of structuredness or optimality, and (c) degree of clarity of requirements definition

(However, a complex may be used in place of monadic functions, since this is a progressive technique.)

Without loss of generality, and for purposes of brevity, we shall assume these factors have been calculated as in chapter 7 (where we discuss them, except 1 and 2, in individual detail). Moreover, our arguments will be developed using basic elementary reasoning.

FUNCTION

The argument for choosing functions comes from direct observations. First, functions are better than the traditional lines of codes. Realizing a function is already comprised of lines of code, functions consequently are more generalized for designing, coding, testing or debugging. The completion of work on a given function is therefore a tangible measure of work done over some time. This is true whether at the design, code, test, or even maintenance phase of a system. Thus, the concept of lines of code is absorbed in that of a function: a function is a distinct composite unit of thought, and a line of code is a derivative of a function.

Secondly, in our calculation we shall count the *sum* or total number of monadic functions worked on. For now, we are not concerned about the particular choice or definition of function—whether global or local or even at level one or two, three, etc. It is mathematically clear that the more functions completed per period, the higher the productivity of a given programmer (designer, analyst, etc.). This explains the reason for the summation. Instead of counting lines of code, we count number of functions; of course, a function may consist of just one or many lines. It is further assumed that these functions are as defined in the visual charts of the software structure diagram for immediate reference. But no assumption is made about the magnitude or size, even in their refinements.

TIME

The duration of time for accomplishing an assignment is a multiplier on the activity (function) performed over that duration. However, unlike automatic processes, this is not a direct multiplier. In most tasks, more work is done over a longer time, assuming constancy of efforts by the participants. Since we are considering our software metric for a particular individual, it would follow that the same programmer would produce more work given more time. But this is not justified by simple observation. The situation is complicated by weighting factors; added to this is the factor of competence. We know a more competent programmer will need less time for a given assignment than a less competent programmer. Since productivity will increase with decreasing time, the time of duration for an assignment will be a divisor for the activities performed. For instance, if the sum of functions is 5 and the time is 4 hours, programmer productivity, ignoring weighting factors for now, becomes:

$$P.P. = \frac{\text{Sum of functions}}{\text{Time}} = \frac{5}{4} = 1.25$$

We observe also that if the time grows infinitely, the assignment will not be completed and productivity tends to zero; this further justifies time as a divisor.

An alternative to this is to choose a weighting factor for programmer competence, perhaps based on seniority or training. Such a factor should be allowed to lie between 0 and 1. Then multiply the sum of the functions by this coefficient and divide by time, i.e.:

$$P.P. = \frac{\text{Sum of functions} \times \text{Coeff. of competence}}{\text{Time}} \qquad (6.3A)$$

It may be necessary in some cases to give consideration to time wasted in waiting, popularly called turnaround time. For a given computing environment, this is not necessary, since this constant cancels out for all programmers. Since turnaround time is an unpredictable happening which may affect some programmers and not others, in a comparison made between two or more centers or two or more programmers this factor becomes significant. Since the effect of turnaround time t_a is to prolong time of completion, T, the effective completion time becomes:

$$T_e = T - t_a \qquad (6.3B)$$

Substituting T_e for T in (6.1), we obtain:

$$P.P. = \frac{\text{Sum of functions} \times \text{Coeff. of competence}}{(T - t_a)} \qquad (6.3C)$$

However, we shall meanwhile drop the coefficient of competence for reasons of eliminating bias and subjectivity.

WEIGHTING FACTORS

There are three of these factors to be discussed.

The first is *degree of complexity*. This is a concept that appears in many books and journals, but without any quantifiable measures or formula or unit. One author defined it as the square of a function. It is obvious that one of the difficulties about the theory of structured programming of software is that few general descriptions can be true of all because no two systems possess the same degree of complexity. A payroll system for one institution may be less sophisticated than that for another institution. An order entry system may be more difficult than a simple queuing system.

More important is that the degree of complexity affects the programmer directly. Just as it is meaningless to consider all modules as of an equal degree of complexity, it is equally meaningless to ignore it in any consideration of measure. To decide where D_c, degree of complexity, varies directly or indirectly, we first note that its most direct effect is to degrade or resist or frustrate productivity. Modules with a lower degree of complexity take less effort than those with a higher degree of complexity. However, complexity is an intrinsic attribute of a system; it is inherent in the definition of the system. Multiplying it by the sum of functions gives a direct picture of the magnitude of resistance to the work (task) to be accomplished. Thus, using the pattern above, by adding just this weight, we obtain:

$$P.P = \frac{\text{Degree of complexity} \times \text{Sum of functions}}{\text{Effective time}} \qquad (6.3D)$$

Degree of structuredness or optimality. Poorly optimized modules will most likely cause prolonged coding and testing—a fact of resistance. It is therefore befitting that we take into consideration this factor of how well a module (system) is structured. Too many systems that claim to be structured are, at best, poorly structured. A poorly structured module may have two or more functions jumbled together, i.e., the functions are ill-refined and therefore lead to poor metrics and productivity. The influence of competence on the part of the designer again shows up here. A well-structured module is a lot easier to work with and vice versa.

A poorly optimized module adversely retards rate of productivity, i.e., the lower the degree of optimality the lower the productivity. On the contrary, the higher the degree of optimality the higher the productivity. We have shown that the variation of the degree of optimality is directly multiplicative. By substituting in the general equation, we obtain

$$P.P. = \frac{\text{Deg. complexity} \times \text{Deg. optimality} \times \text{Sum of functions}}{\text{Effective time}} \qquad (6.3E)$$

The *degree of requirements definition* is a measure of the amount of interference due to changes or managerial intervention. It is traditionally an anamolous norm in the DP environment that specifications are changed during the processes of analysis, design, coding and testing of software. These ad hoc changes cause unpredictable adversities in productivity and are thus an important factor in accounting for programmer productivity. Since this factor has the obvious effect of degrading productivity and therefore penalizing programmer creativity, we notice (from experience) that the more the changes, the more the frustrations and hence the less the productivity. The programmer must therefore be compensated for this adverse imposition. The normalized D_r, degree of requirements, definition is therefore inversely multiplicative. A simple rule to use is to set

$D_r = 1$ if there are no changes during T. If there is one change, set $D_r = \frac{1}{2}$; for two changes, set $D_r = \frac{1}{4}$; in general, for k changes, set $D_r = \frac{1}{2}k$. The compensating effect is seen as k grows, $\frac{1}{2}k$ diminishes and $1/(\frac{1}{2}k)$ increases. Thus, we obtain:

$$\text{P.P.} = \frac{\text{Sum of functions} \times \text{Deg. complexity} \times \text{Deg. optimality}}{\text{Effective time} \times \text{Deg. requirements definition}} \quad (6.3F)$$

In summary, the factors of

T_e, effective time for completing assignment

f_i, (monadic) function in the assignment

D_c, degree of complexity of a system or module

D_s, degree of structuredness of a system or module

D_r, degree of requirements definition of a module

may effectively be used to obtain a numerical measure for programmer productivity, according to the general formula in (6.3F). The simplified notation is:

$$P_p = \frac{\sum\limits_{i=1}^{N} f_i \times D_c \times D_s}{T_e \times D_r} \quad (6.3)$$

$$= \frac{1}{T_e \times D_r}(D_c \times D_s) \sum\limits_{i=1}^{N} f_i \quad (6.3G)$$

$$= P_c \times S_c \times \sum\limits_{i=1}^{N} f_i \quad (6.3H)$$

From (6.3H), the factor

$$P_c = \frac{1}{T_e \times D_r}$$

may be considered as the coefficient of programmer productivity over time, T_e, and the quantity $S_c = D_c \times D_s$ may be considered a characteristic of the system, here called *coefficient of system behavior*. Here is another measure of software equivalence. If two different software have the same characteristic, one can surmise if they are equivalent or not, D_r

being a factor of human influence of interpretation is not an intrinsic attribute of the software behavior. In effect, what we have measured is the coefficient of programmer productivity over time, T_e, given that D_c and D_s are already known from design of the system (not due to the programmer). D_r is the easiest factor to measure. In other words, how many functions, f_i, has the programmer completed over the duration of time? Essentially, the only variables in this formula are time, T_e, and functions, f_i; D_c, D_s, D_r are constant for that system.

We therefore have the two coefficients as follows:

Coefficient of programmer productivity

$$P_c = \frac{1}{T_e \times D_r} \qquad\qquad (6.3I)$$

Coefficient of system behavior

$$S_c = D_c \times D_s \qquad\qquad (6.3J)$$

6.4 SPECIAL FORMULA FOR PROGRAMMER PRODUCTIVITY

◆

So far, all our discussions and developments have revolved around the programmer. We stated earlier that structured programming begins at the analysis phase, and continues through design, coding, testing and integration (and even installation). It is for this reason that the formula (6.3) was called a generalized formula. Consistent with our philosophy of PET, this section will identify and make certain modifications to this equation that will make it applicable to measuring analyst productivity and tester productivity, etc., in case different professionals are performing specific assignments on a software system.

The modifications are as follows: First, we observe that the sum of functions, f_i, is a common variable for all programmers. Therefore, whether we define functions as activities performed or data flow diagrams identified or lines of codes developed, we have a concrete concept of work, though varying according to users. Second, the product $P_c \times S_c$, coefficient of software behavior, gives us a deeper insight into the nature of the software as well as into the programmer's capability to undertake the assignment. With these concepts mastered, we are better able to estimate project completions. Thus, the modifications we make will hinge on the parameter $P_c \times S_c$, especially D_c and D_s since T_e and D_r are dependent on the several characteristic circumstances of individuals.

6.4.1 SYSTEMS ANALYST

The following conditions are directly observable:

1. A design module does not exist, so use the existing factors.
2. The degree of complexity may not be directly reflected in the module's behavior, especially if the analyst has to do another or more analysis on the system.
3. D_s is affected by D_c, but is never known until the designer has completed his or her task.
4. For interview processes with users, D_c and D_s may each be assumed equal to 1.

Any tentative assumptions may be refined by reassembling the proceeds from the interviews or analysis. Such a reassembling will lead to the development above. In other cases, D_c may be measured tentatively from the degree of organization of the analysis. For instance, if the analyst uses data flow diagrams, one can identify the number and organization of the constituent elements (functions) in the flow configuration, in fact, up to their hierarchical expressions using the matrix of their intercommunication. This takes some practice and experience, however. If the analyst uses the functional decomposition approach, the measure for degrees of complexity and optimality become much more simplified. In the final analysis, if D_c is computed as D^*_c we have from equation (6.3I):

$$P_a = P_c \times D^*_c \times \sum_{i=1}^{N} f_i \qquad (6.4.1)$$

6.4.2 SYSTEMS DESIGNER

We begin by making the following observations:

1. Design module does not exist, so use the existing factors.
2. Designer may be reworking an existing design—D_s will be modified.
3. Degree of complexity is inherited from analysis specifications, though to be explored and defined in the finalized design.
4. For interview processes, assume $d_s = 1$.

For the computations for designer productivity we use D_c as determined in (3) above, up until the moment the ultimate degree of complexity can be computed from the structure diagram. In other words, progressive evaluations of designer productivity will use the D_c as determined by the analyst, unless the systems design has been concluded or a design module already exists.

For D_s, degree of structuredness, or PET philosophy dictates we adopt a tentative device as a substitute until the ultimate D_s may be "read" off from a completed software diagram. We proceed as follows: let C_i = number of chain complexes at first level, refined in time T_e; $i = 1, 2, \ldots, N$; $f_{j(i)}$ = number of subcomplexes refined from the C_i's for $j = 1, 2, \ldots, M$. To compensate for the normalization, we take the sum of $f_{j(i)}$ and for each C_i, divide by the product of total number of chain complexes and the total number of subcomplexes to obtain:

$$D^*_s = \frac{\sum_{j}^{\max} f_{j(i)}}{\max i \times \max j} \tag{6.13}$$

EXAMPLE 6.4.2:

Consider for $i = 1,2,3$ and $j = 2,4,6$. Then

$$D^*_s = \frac{\max f_{j(1)} + \max f_{j(2)} + \max f_{j(3)}}{N \times M}$$

$$= \frac{2 + 4 + 6}{3 \times 6} = \frac{12}{18} = \frac{2}{3}$$

Finally, equate D^*_s to D_s. For D^*_s, we have used the condition that at worst,

$$\sum_{j}^{\max} f_{j(i)} \leqslant \max i \times \max j$$

6.5 GENERAL FORMULA FOR EFFECTIVE VALUE OF SOFTWARE

◆

Yourdon [52, p. 28] speculated on a numerical measure for estimating the (degree of) goodness or value of a module. However, he left calculation of the coefficients and or efficiency to the discretion of the manager.

His equation

$$V = a_1 \cdot (\text{testing cost}) + a_2 \cdot (\text{programming costs}) + \ldots$$

$$+ a_2 \cdot (\text{efficiency})$$

to help "judge the effectiveness of the program and, of course, of the programmer who wrote it" raises two questions. First, two or more programmers might have worked on this module, and hence the effectiveness of one programmer becomes more difficult to measure. Second, there is a mixture of cost and non-cost (efficiency) determinants, thus making the equation more difficult.

As a substitute to this useful idea, let's use the following parameters to obtain the effective value of a software, leaving the effective performance of the programmer as measured earlier. The effective value will be related directly to the following costs: analysis, design, coding, compiling, testing. Any other costs may conveniently be included in the generalized equation. Except for coefficient of compilation which is developed below, the respective programmer productivity coefficients of the previous section may be used in the equations below. Finally, we shall add to this set of costs and their corresponding coefficients the critical factor of performance whose coefficient will be the degree of reliability. We propose the following equation for the effective value of a software with p_i as in previous sections and p_6 = degree of reliability:

$$E_s = p_1 \cdot (\text{cost of analysis}) + p_2 \cdot (\text{cost of design})$$

$$= p_3 \cdot (\text{cost of coding}) + p_4 \cdot (\text{cost of compiling})$$

$$= p_5 \cdot (\text{cost of testing}) + p_6 \cdot (\text{performance}) \ldots$$

$$+ p_N \cdot (\text{cost of maintenance})$$

$$= \sum_{i=1}^{N} p_i C_i \tag{6.5A}$$

For p_3, coefficient of compilation for programmer productivity, we proceed as follows: There are two significant variables: number of compiles and number of diagnostic messages. Since the latter affects the former directly, we divide the number of compiles by the degrading factor, number of diagnostic messages. Therefore, we obtain:

$$P_3 = \frac{N}{M} = \frac{\text{No. of compiles}}{\text{Sum of diagnostic messages}} \tag{6.5B}$$

The sixth factor in E_s, performance, has a cost value. It is the cost of memory size. Since performance is defined as a variable of core size (obtained from compiled versions) and execution time, it follows that time of execution is affected by core size. Alternatively, both costs (core size and execution time) may be summed. Core size may be taken from the linkage output listing.

The uses of this general cost-effective equation may also be theoretical at the initial definition of the phases of a project and stretching over a whole life cycle. The traditional practice of making arbitrary cost estimations may now be replaced by a more powerful tool. In practice, arbitrary coefficients may be substituted and tentative costs assigned; then by more trial and error, even by a simulated computer run, an approximate feasible cost estimation may be arrived at. Such a simulation may use the simplex method of linear programming (programming = planning [24, p. 127]). Arbitrary coefficients may be assumed to satisfy a set of cost constraints in order to solve for a minimal (optimal) cost. Here is a tool no manager can afford to miss.

EXAMPLE 6.5A: Simplex Model

Since cost is an overriding concern, we might impose some cost constraints on a set of program coefficients in order to minimize cost. We choose coefficients over the range $0 < p_i \leqslant 1$.

For $i = 1$: $p_{11}, p_{12}, p_{13}, \ldots, p_{1N}$; and $\sum_j p_{ij}C_j \leqslant B_1$

For $i = 2$: $p_{21}, p_{22}, p_{23}, \ldots, p_{2N}$; and $\sum_j p_{2j}C_j \leqslant B_2$

For $i = 3$: ...   ...

For $i = 4$: ...   ...

...

For $i = k$: $p_{k1}, p_{k2}, p_{k3}, \ldots, p_{kN}$ and $\sum p_{kN}C_j \leqslant B_N$

And, in general, we obtain the rows of inequalities:

$$\sum_i \sum_j p_{ij}C_j \leqslant B_i \tag{6.5}$$

Finally, apply the simplex algorithm to (6.5) to obtain a set of minimal costs given a set of constraints of budget. The simplex algorithm can be found in some mathematical packages. It is very useful for systems analysis, and, in particular, for feasibility analysis. According to its methods, we can transform (6.5) into a set of simultaneous equations by adding slack variables to each row to obtain:

$$\sum_j p_{ij}C_j + C_{j+1} = B_i; \quad \text{which leads to}$$

$$\sum_i \sum_j p_{iv}C_v = \sum_i B_i \qquad\qquad (6.5C)$$

EXAMPLE 6.5B: Minimize Cost

The available time for a software phase which might limit productivity is as in Table I and the number of programmer hours required per week is as in Table II. The systems director determines that overtime assignments for phases 1 and 2 exceed the maximum production rate, and that for phase 3 is 20 hours. Suppose further that the proposed pay rates for programming technicians are $25, $35, and $45. Our objective function is

$$Z = 25C_1 + 35C_2 + 45C_3$$

subject to (using the variables in the tables):

$$3C_1 + 2C_2 + 8C_3 \leqslant 50$$

$$2C_2 + 4C_3 \leqslant 100$$

$$C_1 + 2C_3 \leqslant 200$$

$$C_3 \leqslant 20$$

TABLE I

Phase	Avail. time
Analysis	50
Design	100
Testing	200

TABLE II

Phase	Programmer hrs. per week		
	P_1	P_2	P_3
Analysis	3	2	8
Design		3	4
Testing	1		2

EXERCISE 6.5C: Maximize Profit

The management of a software house uses 3 service bureaus for production. The output of each bureau is limited by an assignment of production (user) time (Table I). The productivity of management is also limited by their available number of programmers (Table II). The planned production according to expected profit and duration to completion are as shown in Table II. In order to maintain a uniform load of productivity among the bureaus, management decides for an equal percentage of hours at the bureaus. Management would like to maximize its profits (sales).

TABLE I

Bureau	Usable hrs	Available pgrmr hours*
1	400	1500
2	600	2000
3	300	900

*Not necessarily per bureau

TABLE II

Max. hrs.	Pgrmr. hrs/project	Profit ($)
700	5	400
800	4	300
800	3	100

The decision variables are X_{ij}; for $i = 1,2,3$, the bureaus; and $j = 1, 2, 3$, the production hours (actual). Therefore, the objective is to maximize

$$Z = 400(X_{11} + X_{21} + X_{31}) + 300(X_{12} + X_{22} + X_{32})$$

$$+ 100(X_{13} + X_{23} + X_{33})$$

subject to the following constraints:

Usable hours:

$$X_{11} + X_{12} + X_{13} \leqslant 400$$

$$X_{21} + X_{22} + X_{23} \leqslant 600$$

$$X_{31} + X_{32} + X_{33} \leqslant 300$$

Available programmer hours:

$$5X_{11} + 4X_{12} + 3X_{13} \leqslant 1500$$

$$5X_{21} + 4X_{22} + 3X_{23} \leqslant 2000$$

$$5X_{31} + 4X_{32} + 3X_{33} \leqslant 900$$

Production hours (maximum):

$$X_{11} + X_{21} + X_{31} \leqslant 700$$

$$X_{12} + X_{22} + X_{33} \leqslant 800$$

$$X_{13} + X_{23} + X_{33} \leqslant 300$$

Uniform workload:

$$\frac{X_{11} + X_{12} + X_{13}}{400} = \frac{X_{21} + X_{22} + X_{23}}{600}$$

$$\frac{X_{21} + X_{22} + X_{23}}{600} = \frac{X_{31} + X_{32} + X_{33}}{300}$$

$$\frac{X_{11} + X_{12} + X_{13}}{400} = \frac{X_{31} + X_{32} + X_{33}}{300}$$

Although there is no empirical investigation, one may seriously suspect that a major cause of failure of economic (and other) models stems from the use of inadequate or inadequately computed coefficients. Some organizations have preferred rules and standards which the analyst must use, thus denying him or her an openminded search for appropriate methodology. The selection of a methodology deserves the same careful attention as the actual procedural computation from a model. Chapin [11] summarizes the conventional techniques of economic modelling:

1. Net present value: plots costs and benefits over time; recognizes quantitative flow of money. This method requires unambiguous decision on time interval or value of money; it may involve heavy calculations.

2. Cost and benefit ratio: emphasizes the cost of life cycle and balance of benefit. Its main illusion is that ratios do not reveal the actual quantities.

3. Time to pay off: determines how long it takes to receive payment on a given investment. This uncertainty carries the penalty of unknown return time, and interaction with an organization's revenue.

4. Breakeven analysis: based on the distinction between the variability of factors—costs, benefits; has limited application.

5. Least cost: based on the philosophy that cost is easier to observe, measure and control. Ignores variables of benefit and hence lacks sound foundation.

6.6 GENERAL FORMULA FOR ACCEPTANCE TEST OF SOFTWARE

From the user's perspective, the value of the end-product is best measured in terms of degree of correctness and of reliability. Of the five coefficients, these are the only two that directly affect and ultimately determine the quality assurance of their product. The problems of adaptability are all directly influenced and affected by these also. Since the two coefficients are probability measures, here is a favorable place to design hypothesis experiments. Even though the degree of requirements definition is generated by the user, its impacts are reflected on programmer productivity and not on user productivity. It may be tempting, also, to consider degree of complexity here, but its impacts are already absorbed in degree of correctness and of reliability.

Using the same arguments of variational factors, the two coefficients can either degrade or uplift user productivity. They are therefore jointly multiplicative, and we obtain:

Coefficient of quality assurance, $Q_c, = D_T \times D_R$ (6.6)

Since PET uses tentative values during certain phases of development only to have *actual* values substituted at the right time, and since at the end of software testing these values are known, it is believed that Q_c will carry actual values of D_T and D_R. This makes it practical.

EXERCISES

6.1 List current (up to 1982) methods of measuring programming productivity.

6.2 Identify and describe any factors that you consider pertinent to productivity. List them in their order of importance and examine their interactions.

6.3 Use Figures 5.3.2B and 5.4.2B to verify Equation 6.3, assuming any arbitrary time, T_e.

6.4 Use Figures 5.3.2B and 5.4.2B to verify equation 6.4.1, taking into consideration interview processes all the way through analysis specification.

6.5 Design a simplex algorithm for formula 6.5A, and show how it can be used to monitor or manage software development in terms of costs and manpower planning. Compute actual coefficients of productivity by substituting actual cost factors and some approximate costs. This solution will serve as a useful check for future control of production.

Chapter 7

SIGNIFICANT DETERMINANTS for SOFTWARE METRICS

Science, like all other disciplines, has its own aura of revolution. It is a quiet revolution, less talked about than ordinary news dispatches [30, p. xiii]. The heart of this revolution is ideology or polemics. Spoken or written, each idea or postulate is accepted or rejected after deliberation on the basis of its logical validity. A new theory or principle is then born; that which is "imperfect" is rejected for that which is "perfect".

7.1 DEGREE OF SOFTWARE COMPLEXITY

In the past, many writers and managers have relied on number of lines of code to judge the degree of complexity of a system or module. According to this approach, a module with 4000 lines of code is more complicated than one with 2000 or even 1000 lines of code. This is based on the falacy of man's habitual bias for size—the bigger the more perplexing. There is no rational vindication for this judgment. As mentioned earlier, two programmers are likely to code a given module with a different number of lines, depending on their experience and proficiency. The developments here will discard this approach as unempirical.

The degree of complexity of softwares is a *real attribute* of the system. It is therefore not measurable by the methods of probability distribution functions which are useful for apparent observed values. Rather it is measured with the actual observable attributes of the system. The effects of complexity are significant in determining such software characteristics as independence of modules (functions). We shall explore the problem of degree of complexity of a system (based on the software structure) (5.3.2) in terms of the impact contributed by the following:

1. Complexes (for modules) or modules (for systems)—levels of refinement or decomposition of a complex
2. Expanse—number of chains of complexes in the system
3. Degree or order of refinement of a chain
4. Order of selection structures

The role of these factors in determining complexity can be introduced as follows (Figure 7.1):

1. For a system, we shall consider the modules as functions (of the system). In other words, modules "look" like or are equivalent to functions in a software structure design. Clearly, non-function oriented designs lack this property. This is the great advantage in advocating the use of functions in studying systems. We can shift our frame of reference easily from modules (of a system) to functions (of a module) without loss of generality. In essence, a module is a "bundle" of functions, which are eventually individually refinable, and a system is the direct sum of modules.

The *level* of refinement of a complex is expressed by a whole number, L, say. As L increases as in a large tree diagram, the complexity of the representative software grows. If there are L levels, the monads at level $L + 1$ have dimensions $= L$. As each subcomplex further refines, the complexity of a system grows almost "exponentially."

2. Considering the expanse of growth (i.e., how many subcomplexes are obtainable from a given complex), we observe another "orthogonal" growth. It is natural to measure this factor at the first level, since each complex here generates a chain of subcomplexes. The total number of complexes at this level will be called the order of the expanse denoted by the whole number, E.

Thus, the combined growth factor is a measure of level and expanse:

$$G = L \times E \tag{7.1A}$$

3. The concept and terminology of monad is very helpful here in explaining what follows. Its potential application is a direct factor for understanding the concept of the degree of complexity. In the process of refinement, there is an "unconscious" count of the number of refinements a particular complex (module or function) has undergone. The deeper the refinement, the higher the count; see, for example, Figure 5.4.2B.

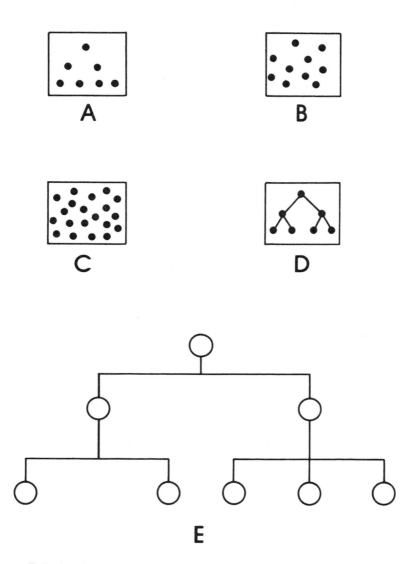

Figure 7.1 An intuitive approach to understanding software metrics, especially *complexity* and *reliability*. A) Sparce distribution of nodes. Each node represents a function/activity type. B) Denser distribution of nodes. C) Increasing density of nodes reflects the growing complexity of the structure. D) A simple topologically connected set of nodes of Figure 7.1A; this is our first concept of counting refinements. E) Formal structural annotation of Figure 7.1D.

This count is called the dimension of a monad. Semantically, the number of subcomplexes refined out of a given complex is analogous to the number of degrees of freedom, hence, dimension. For our metric, we note that *each* chain or sequence (originating at level 1) eventually terminates at some level with a set of monads, M_j, say. Consequently, we obtain for all terminating monads.

$$M = \sum_{j=1}^{Z} M(E_j). \tag{7.1B}$$

where $M(E_j)$ is the total number of monads of the E_j-th chain in the system. The presence of these chains of complexes suggests that the popular terminology top-down is somewhat inaccurate. It assumes we are working down a single complex and ignores the other chains.

4. The select structure has two types: nesting and casing. Of the set of nested-IFs, there is exactly one that has the most nesting of IF-constructs. The number of distinct IF-predicates in a given nesting will be called the degree of that nested-IF construct. This number is equated to S_1.

Similarly, the number of action statements (those constructed with a THEN or ELSE condition) will be called the degree of the case or select statement and equated to S_2. The combined degrees of these will add up to S:

$$S = S_1 + S_2 \tag{7.1C}$$

What is here introduced as order of selection structure is a general language for a number of alternative actions. Some systems have such structures that every module belonging to a certain node must be processed; others have their modules processed according to data conditions. These data conditions may generate further subordinate data conditions—giving rise to say, K, nested-IF structures. These systems are evidently more complex than the former. This is the reason for highlighting this attribute of systems as contributing to complexity. The average programmer knows the complicating effect of deep nesting IFs. However, since both S_1 and S_2 contribute, under our new generalized tree structure, to deeper levels, the effect of S is already absorbed in G.

In conclusion, our argument about dimension already demonstrates that the deeper the levels, the higher the dimension. This reflects directly on the amount of work to be done and, hence, the degree of complexity. The second argument is that the degree of refinement actually reveals the complexity of the individual functions constituting the system. The third argument is that the order of expanse contributes to the degree of complexity; it alone defines how many chains are present in the system. Therefore, the three factors add up multiplicatively to the degree of complexity.

We shall use these *three* attributes to design by structure to measure the degree of complexity, D_c, of software, obtaining:

IMAGE of complexity, $\hat{D}_c = L \times E \times M$

$$= L \times E \times \sum_{j=1}^{E} M(E_j) \qquad (7.1)$$

Caution: $M(E_j)$ is a *single* number; it is the number of monads in the J-th chain in the count of expanse. Strictly, we have

$$M(E_j) = \sum_{k=1}^{M} m_k \text{ and } E = \sum_{j=1}^{N} E_j \qquad (7.1D)$$

as the total number of monads of the E_j-th chain.

It is clear from this definition that D_c is greater than zero, demonstrating that every software system has some degree of complexity. In summary, D_c depends on the following factors:

L, the level of refinement—dimension of the lowest monad

E, the expanse or order of level-orthogonal growth

M, the total number of monads, emphasizing the importance and necessity for complete refinement

Without resorting to rigorous mathematical transformation, in order to normalize $\hat{D}_c$ we invoke the principle of complexity that the more the number of functions present, the more the difficulty of efficiently completing the task. We therefore normalize (7.1) by taking a simple reciprocal to obtain:

$$D_c = f_{\text{norm}}(\hat{D}_c) = 1/\hat{D}_c \qquad (7.1E)$$

For computational convenience, we will impose the condition that for any chain, the multiplicative factor $M(E_j)$ is at least 1.

EXAMPLE 7.1A: Single-level System

Suppose a given system has all its modules/functions refined completely at level 1. Further, let us assume it has, depending on the type of system, the following number of refinements: 1, 2, 3, 6, and 8. We now compute D_c for each case ($M = 1$ for each case):

By (7.1) for $E = 1, \hat{D}_c = 1; \rightarrow D_c = 1$ by (7.3)

$$E = 2, \hat{D}_c = 2; \rightarrow D_c = 0.5000$$

$$E = 3, \hat{D}_c = 3; \rightarrow D_c = 0.3333$$

$$E = 6, \hat{D}_c = 6; \rightarrow D_c = 0.1667$$

$$E = 8, \hat{D}_c = 8; \rightarrow D_c = 0.1250$$

It is clear that D_c looks very bad as more functions are encountered or added to the system.

EXAMPLE 7.1B: Multilevel System

Suppose a system has $L = 6$, $E = 4$, and $M(E_j) = 3, 6, 4$, and 2 for $j = 1, 2, 3, 4$, respectively. Then our computation leads to:

$$\hat{D}_c = L \times E \times \sum_{j=1}^{4} M(E_j) = 6 \times 4 \times (3 + 6 + 4 + 2) = 24 \times 15 = 360$$

By (7.1E), $D_c = 1/360 = 0.0028$.

EXAMPLE 7.1C: D_c for Figure 3.3A

The system shown in this figure has $L = 3$; $E = 5$; the monads are distributed as follows: $M(E_1) = 2$; $M(E_2) = 3$; $M(E_3) = 4$; $M(E_4) = 4$; and $M(E_5) = 1$, by imposition rule. Note, we have used 4 for $M(E_4)$ rather than 5 because by definition the lowest level monads are used. We therefore obtain:

$$\hat{D}_c = 3 \times 5 \times (2 + 3 + 4 + 4) \ (M(E_5) \text{ has its only monad at level 1 counted in } E)$$

$$= 15 \times 13 = 195$$

By (7.1E), D_c becomes $1/195 = 0.0051$.

In conclusion, what we have is: If N is the total number of functions/ modules in a system, then

$$\text{Lim } D_c = \underset{N \to \infty}{\text{Lim}} \frac{1}{N} \to 0 \qquad\qquad (7.1\text{F})$$

This is the meaning of complexity. Here is one elementary principle for numerically comparing software systems. It was not advisable to use N directly since we must include the effects of refinement and chains in the computation.

Thus, we have this definition: The degree of software complexity is a measure of software structure in terms of the interconnections of its functions/components.

7.2 DEGREE OF SOFTWARE STRUCTUREDNESS OR OPTIMALITY

◆

The degree of structuredness of a system (module) is directly influenced by the degree of complexity. But since the latter is inherited from the absoluteness of the nature of the system while the former is affected by the competence of the designer, we shall treat D_s, degree of optimality, as a separate entity. The following factors will be examined for computing D_s:

1. Independence of modules/functions

2. Irreducibility of modules/functions

3. Intermodule (Interfunction) communication: (a) super- and subordinate linkage, (b) single-entry/single-exit property

The concept of independence of modules or functions is difficult to define. Yourdon [11, p. 96-7] considers independence of a module with respect to effects being inflicted on one module by changes in another as regards:

1. Program logic

2. Shared arguments of data

3. Internal variables

4. Data base

5. Flowpath or control path

Myers [37, p. 23], on the other hand, gives the following criteria: Comprehensibility, Maintainability, Reusability and Extendability.

However, the author has a simpler approach. Only two structures are necessary: data structures—set of all input/output; and flow structures—set of all functions. Independence then reduces to inequivalence of input/output data sets and of functions. Since this factor is still too cumbersome to measure, we can obtain a "quick" simplified measure of independence by using the set of data structures. We define two modules as *relatively* (under our new prescription) independent if their data sets are unique (i.e., nonintersecting). (A full independence will include function uniqueness, in addition to data set uniqueness.) Consequently, independence is not true for modules that CALL or are CALLED by others.

In order to remove a minor complication introduced by this definition, if N = number of intersecting data sets between any pair of modules (functions), we assign a value of $1/(N+1)$ for independence. Consequently, for the subfactor of independence, we have

$$\text{Degree of independence, } d_i = 1/(N+1) \tag{7.2A}$$

and the combined product of independence measures becomes

$$\hat{d} = \prod_{i=1}^{L} d_i \tag{7.2B}$$

where L = number of pairs of modules affected. Thus:

$$d = \begin{array}{ll} 1 & \text{if independence holds} \\ \hat{d} & \text{if independence fails} \end{array}$$

A very significant contributing factor to optimality is the degree of refinement of functions or modules. This is where designers falter. Ultimately, all the functions that form the leaves of a software structure diagram must consist of irreducible functions, i.e., functions that cannot be refined anymore. The equivalent to this in the case of a system is irreducible modules. The only time to vindicate a designer for irreducibility is at coding time, but for ad hoc computations of productivity, we shall introduce a probability factor. This probability factor will be a 50/50 or so chance that each monad at the lowest level is irreducible, that is, has been minimally refined. By taking the expected value, the numeric value for irreducibility becomes

$$\hat{r} = E[r] = \sum_{i=1}^{M} M_i \times P_i \tag{7.2C}$$

where M_i = the i-th leaf or monad, and p_i = probability of refinement for the i-th leaf. To normalize, we divide $\hat{r}$ by ΣM_i to obtain

$$r = \frac{\hat{r}}{\Sigma M_i} \tag{7.2D}$$

Finally, our hierarchy structure requires that each module at any level be accessed *only* by its immediate father node. A violation of this father-son relationship (true for EXIT-IN-THE-MIDDLE style) will degrade system optimality. Both subfactors of (3) are clearly related and will be treated simply as communicational factors.

Numerically, if each module has communicational single-entry/single-exit, we assign a value of 1 to the factor of intermodule (or interfunction) communication. In general, for the M_i-th node

$$C_i = \frac{1}{J} \times \frac{1}{K} \tag{7.2E}$$

where j = number of distinct entries into M_i and k = number of distinct exits from M_i. It is clear that if $j = k = 1$, $C_i = 1$. The combined product effect of the C_i's will give for the whole system or module

$$C = \prod_{i=1}^{M} C_i \tag{7.2F}$$

where M = total number of pairs of intercommunicating modules, and multiplication indicates the degree of degradation caused by C. In practice, J and K count how many modules (functions) access a given module (function). Summarizing, in consistency with our idea of normalization, and realizing that these subfactors of structuredness (optimality) contribute to the degradation of a system, we take the product of D_S as

$$D_s = c \times d \times r \tag{7.2}$$

where:

c = communication subfactor (single-entry/single-exit)

d = dependency subfactor (of commonality of data/function)

r = refinement subfactor (trace of hierarchy)

Since each subfactor is greater than zero and less than or equal to 1, D_s is already normalized to

$$0 < D_s \leqslant 1$$

Observe that the formula for D_s turns out to be the most complicated of all the significant determinants discussed in this book. It is, therefore, no triviality that designing a system needs all the care, observation, analysis and abstraction there is. Many existing systems that are said to be "structured" are really short of that goal.

EXAMPLE 7.2A

Let us compute D_s for the system of Figure 3.3A. There are three parts to the solution:

1. Degree of independence: The labels may not reveal the actual degree, but the coding (data structure and function paths) will. However, for simplicity, assume independence holds at level 1. Detailed independence may be computed for all the functions (modules) in the system. For most practical purposes, we consider only the modules at level 1. Thus, $d = 1$.

2. Degree of refinement: Consider *two* cases: (a) Actual/projected probability of refinement = 1; (b) computed/projected probability of refinement = 0.5.

3. Trace of hierarchy: If coding conforms to design specification based on the design diagram, in which case all single-entries and single-exits are into and out of individual functions being processed and there are no EXIT-IN-THE-MIDDLEs, then, $C = 1$ from (7.2F).

Using (7.2) and combining number (1) and number (2a) above and trace of hierarchy, we obtain:

$$D_s = c \times d \times r = 1$$

Combining (1), (2b) and 3, we obtain:

$$D_s = c \times d \times r = 1 \times 0.5 \times 1 = 0.5$$

Note that from (7.2D),

$$r = \frac{(1 \times .5) + (2 \times .5) + (3 \times .5) + (4 \times .5) + (5 \times .5)}{15}$$

$$= \frac{7.5}{15} = 0.5.$$

We can extend this example by the following observations: If C in (2) is further degraded by having j or k greater than 1, D_s will wind up with a worse value; on the other hand, if the probabilities are improved, D_s will, too, and vice versa.

In particular, consider a single EXIT-IN-THE-MIDDLE construct in some single function. This calculates:

$$C = \frac{1}{k} \times \frac{1}{j} = 1 \times \frac{1}{2} = 0.5$$

There are *two* exits. Further, two such constructs in the *same* function will calcuate to:

$$C = 1 \times \frac{1}{3} = 0.33$$

There are *three* exits. This demonstrates the degradation effect of communication on structure.

7.3 DEGREE OF SOFTWARE REQUIREMENTS DEFINITION

◆

The degree of requirements definition will be presented here as the least troublesome of the measure factors, although in practice it is one of the most troublesome during development. The reason for introducing this factor is because it is a well-known controversial obstacle.

In general, there is changing, adding, deleting and modifying of instructions, functions and requests during the process phases of a system. The average professional knows fully well the traumatic impact this activity may have on the programmer and, consequently, on productivity, usually lagging for days or even months. It is therefore vital to computing programmer productivity. It is the main focus of communication between management and technicians. Sometimes, the deficiency is introduced by the weakness of management to understand fully or conceive effectively what they want to do. Other times, the complication is introduced because of some misunderstanding on the part of technical personnel of what is documented in writing or narrated verbally. Or still, the error could be one of inherent logical complexity. Each change made to the system influences cost and time of production directly. Its significance cannot be ignored.

The one difficulty to be encountered here is determining the weight of the changes in the requirements definition. It is true that such a change may force several changes on our computations for degrees of complexity and optimality. In chapter 6 we postulated D_r as a diminishing factor on programmer productivity and as inversely multiplicative. One elementary measure for D_r is to count the number of interferences or changes made to a system up to the moment of measuring programmer productivity. This approach will simultaneously normalize our factor. Thus, for the ideal case where there is no change, we simply set $D_r = 1$. For $1, 2, 3$, or 4 changes, $D_r = \frac{1}{2}, \frac{1}{3}, \frac{1}{4}$, or $\frac{1}{5}$, respectively. It is clear that as the number of changes increases, D_r degrades productivity, making it tend to zero. Thus, in general, we define

$$D_r = 1/(N + 1) \tag{7.3}$$

where N = total number of interferences over T, the duration time of the assignment. When we divide with this, we can then see the compensating effect of D_r on the programmer's efforts.

It may be necessary to compute D_r to a higher accuracy. This will require multiplying D_r (as computed above) by some coefficient that will reflect the gravity or criticality of the interference. This is natural because some interferences are more agonizing to the programmer, whereas others may be minor and amenable to the system. One suggestion for determining this coefficient is to use the number of functions or instructions in the given interference; another is to let it be governed by the phase of the development—analysis, design, coding, compiling, testing, etc.—or even a combination of both. The actual numerical value assigned to this coefficient will depend on the experience of the computing organization; otherwise, it may be determined arbitrarily. It is important to ensure that this computed coefficient be normalized to lie between 0 and 1, but never equal to 0.

Therefore, our measure for D_r becomes:

$$D_r = \frac{C}{(N+1)} \tag{7.3A}$$

where C = coefficient of interference, and N = total number of interferences.

One approach to computing C may be as follows: For changes made for

$$
\begin{aligned}
\text{Analysis} &: & C_a &= \frac{1}{4} \\
\text{Design} &: & C_d &= \frac{1}{4} \\
\text{Coding} &: & C_c &= \frac{1}{8} \\
\text{Compiling} &: & C_m &= \frac{1}{8} \\
\text{Testing} &: & C_t &= \frac{1}{4}
\end{aligned}
$$

If further accuracy is desired, multiply the appropriate one of these with C_k, where C_k = the reciprocal number of instructions affected, to obtain:

$$C = C_k \times C_i \tag{7.3B}$$

where C_i = one of C_a, C_d, C_c, C_m, or C_t. Otherwise use the C_k for (7.3A).

EXAMPLE 7.3A: D_r/Figure 3.3A

Suppose a software development had 25 changes made to its requirements definition at various points during its life cycle as follows:

Changes	Stage
5	systems analysis
8	design analysis
5	coding
3	compilation
4	testing

Since these are actual occurrences and have the same units, we can sum the combined effects using (7.13) and (7.14) as follows:

$$D_r = \frac{1}{4}/6 + (\frac{1}{4}/9) + (\frac{1}{8}/6) + (\frac{1}{8}/4) + (\frac{1}{4}/5)$$

$$= \frac{1}{4} \left(\left(\frac{1}{6} + \frac{1}{9} + \frac{1}{4} + \frac{1}{5} \right) \right) + \frac{1}{6 \times 8}$$

$$= \frac{1}{4} \left(\frac{30 + 20 + 45 + 36 + 15}{12 \times 3 \times 5} \right) = \frac{146}{48 \times 15} = 0.2027$$

EXAMPLE 7.3B:

Suppose now that there were 5 changes, one for each phase, instead of the previous 25. A new calculation gives the following:

$$D_r = \frac{1}{2} \left(\frac{1}{4} + \frac{1}{4} + \frac{1}{8} + \frac{1}{8} + \frac{1}{4} \right) = 0.50$$

These two contrasting examples clearly illustrate the gravity of the impact of changes during software development.

7.4 DEGREE OF SOFTWARE CORRECTNESS

◆

The major difficulty in computing the degree of correctness of software is the impossibility of totally predicting the existence of bugs. From the principles of mathematics, a rational computation of correctness must follow the theory of probability—we seek the probability that a given software gives correct results. Correctness, unlike complexity, is an apparent attribute of software.

However, Dijkstra's warning is valid. We must concern ourselves with what "program structures" can lead to correct results. From among the class of program structures, we choose the equivalent class of well-structured systems and examine them for correctness.

The theory of (degree of) correctness reinforces the functional approach to programming. Furthermore, functions are amenable to refinement and it is easier to test correctness with the monads. Then we can apply Boolean predicate tests to validate these functions. This is done along the chain of hierarchies. The reader is reminded that *hierarchy* is defined for both function and data structures. Of the two, function hierarchy lends greater speed and convenience for Boolean tests of validity.

This decision eliminates spaghetti systems and ill-structured systems (modules). We will compute formulae that this selected class of systems have the potential of satisfying their requirements definitions. The methods of testing are concerned that the path predicates are satisfied or correct; however, the probability of being correct is not concerned with the "causes" of correctness or failure. Therefore, we direct our efforts to the actual "performance" of the constituent functions, and thus the "ingredients" of our measure will be the program complexes.

In practice, the computation of degree of correctness poses this problem: Suppose we are given a function, F, composing a monad. Demonstrate that F is computed correctly. Clearly, we must show that F satisfies its domain and range. In addition, given any elements outside its domain, F fails to map them into its range; in debugging jargon, F "kicks" them out, causing the execution to terminate or display some error message. The method of accomplishing this demonstration is to construct Boolean predicates on F to conclude the truth or falsity of the Boolean statement. This might be cumbersome considering the size of the domain of F. The following approach is consistent with PET.

Correctness is concerned with the *quality* of a product—a product is either defective or effective. Therefore, the quality of a product is a Bernoulli Random Variable [26, p. 37]. In dealing with the derived nodes of a software structure, a defective node may propagate its bug to another node. Correctness is thus contagious. This information is, however, useful for debugging; the degree of correctness is not concerned with the sources of bugs. For a node at level k, say, to be correct, all its ancestor nodes at level $j < k$ must be correct. At this point in time, we have no knowledge of the correctness of the successor nodes. This suggests one possible model is the elementary Markov Dependent Bernoulli Trials; its sample space has only two events—success and failure. For a recursive computation in terms of difference equation, see [42, p. 130-4].

However, since we do not know the individual (nodes) probabilities, we shall here investigate another model. We shall assign arbitrary probabilities to the nodes at level 1. If a node, A_k, refines into N nodes (twins), we assign:

Probability (error in any member of twins) = $1/N$.

By the contagious character, its actual probability = $P(A_k)*1/N$.

In order to increase precision, we shall consider each chain as independent random variable. For each chain, C_i, using the notions of set theory [42, p. 15, 22 (Ex. 5.3)] we have

$$P_i = P \text{ (all nodes are correct)} = P(\bigcap_{i=1} A_i)$$

$$= 1 - P(\cup \overline{A}_j) \text{ where } \overline{A}_j \text{ is the complement of } A_j$$

$$= 1 - \sum_{j-1} p(\overline{A}_j) + \sum_{m=1}^{L} \left(\prod_{j=2}^{m(k)} \cap A_j \right) \qquad (7.18A)$$

To determine the degree of correctness for the whole system, we compute the expected value

$$C = E \text{ (systems is correct)} = \sum_{i=1}^{n} i*p_i \qquad (7.4B)$$

Finally, we may normalize this by dividing by

$$\sum_{k=1}^{n} K = N(N + 1)/2 \text{ to obtain}$$

$$D_T = 2C/N(N + 1) \qquad (7.4C)$$

EXAMPLE 7.4A: $[D_T/\text{Fig. 3.3A}]$:

Let us now calculate the degree of correctness for the system of Figure 3.3A. For the nodes at level 1, considering the chains as independent random variables (alternatively, we may consider the whole system as a single random variable, in which case we must not normalize again), we assign the following probabilities:

$p\langle 0,1\rangle = .75; p\langle 0,2\rangle = .50; p\langle 0,3\rangle = .40; p\langle 0,4\rangle = .40; p\langle 0,5\rangle = .95$

$C_1:p_1 = 1 - .25\epsilon_1 = 1 - .25[1 - 1/2 + 1/2 - (1/2)^2]$
$\qquad = 1 - .25[1 - 1 + 1/4] = 0.9375.$

$C_2:p_2 = 1 - .5\epsilon_2 = 1 - .5 \ [1 - (1/3)*3 - (1/3)^2*2 - (1/3)^3]$
$\qquad = 1 - .5 \ [1 - 1 - 2/9 - 1/27] = .08714.$

$C_3:p_3 = 11 - .6\epsilon_3 = 1 - .6 \ [1 - (1/4)*4 - (1/4)^2*6 - (1/4)^3*4 - (1/4)^4]$
$\qquad = 1 - .6 \ [1 - 1 - 6/16 - 1/16 - 1/256]$
$\qquad = 1 - .6 \ [1 - (256 - 96 - 16 - 1)/256] = 0.7349$

$C_4:p_4 = 1 - .6\epsilon_4 = 1 - .6 \ [1 - 1/2 - 1/4 - (1/2)e_{42}]$
$\qquad = 1 - .6 \ [1 - 1/4 - (1/2)[1 - 1 - 113/256]]$

where ϵ_{42} = probability for the nodes $\langle 0,4,2,1\rangle$ thru $\langle 0,4,2,4\rangle$ computed as above for $\langle 0,3,1\rangle$ thru $\langle 0.3,4\rangle$;

$\qquad = 0.6826.$
$C_5:p_5 = 1 - .05 = 0.95.$
From (7.4C), $D_T = [1\text{x}.9375 + 2\text{x}.8714 + 3\text{x}.7349 + 4\text{x}.6826 + 5\text{x}.95]/15$
$\qquad = 0.8244$

Let us summarize by making two comments. Firstly, consistent with (3) there are two major approaches to demonstrating software correctness. One is called *testing*, which applies to executing the software with test data and then determining how best the output correlates with the requirements definition. The other is *verification*, which resorts to rigorous mathematical proof that the program is consistent with its specification. The latter does not require executing the program. In contrast, the former does not require a knowledge of the program's structure [56, p. 113].

Secondly, there is a distinction between correctness and reliability of software. Correctness is concerned with the consistency between the results of a program's code and the program's requirements specification. A software (or component of) is correct if and only if this consistency holds. On the other hand, reliability is a statistic of software behavior in its environment. This statistic measures the degree of confidence we have in the performance of the software which is a reflection of its design. According to Wolf [56, p. 105], correctness is a necessary condition for reliability.

7.5 DEGREE OF
SOFTWARE RELIABILITY

♦

An ideal software, one that never failed throughout its production life, is said to have 100% reliability. All other cases exhibit a reliability of less than 100%. Reliability is a pathological effect of design. Our task here is to examine how to measure reliability. In this pursuit, we will focus on failures, failure times and failure frequencies.

Many of the theories used to describe reliability of software systems turn out to be applicable to hardware, and not even suitable to software. This is because a suitable structure to describe software does not as yet exist. The components of hardware deteriorate with time, in contrast to those of software which are fixed and "permanent." Besides, the interactions between the components of software are much more complicated than those of hardware [56, p. 99]. These observations suggest we look elsewhere for the study of software reliability, away from those hitherto proposed. In particular, software reliability encompasses two broad categories of thought—availability and integrity [56, p. 108]. *Availability* has to do with the environment of application, and *integrity* with correctness of the software. The reliability of a software system depends on both the total number of design errors and the environment of use.

The theory of reliability, as it is used throughout engineering, has its origins in mathematical statistics. If a given product or item has a time to failure, T, then its reliability for a fixed length of time t is defined to be

$$R(t) = P(T > t) = 1 - F_T(t) = e^{-\lambda t} \tag{7.5A}$$

where λ is the parameter for the exponential random variable T, and $P(T > t)$ is the probability that the random variable T is greater than t; e is the well-known mathematical exponential function. λ may be estimated by any of several experimental methods, one of which is to take N identical products as a sample, and wait until all N products have failed (time to failure t is known). Two prominent methods in this philosophy are:

1. Mean Time Between Errors (MTBE): This is the *statistical* mean (as opposed to average) of operating times between interruptions.
2. Mean Time To Repair (MTTR): This is the *statistical* mean of the length of time to correct a bug.

These methods, though statistically sound, are probably not adequate for measuring software reliability. It is very important to realize that statistics may yield ineffective results depending on the type of sample taken, on the method of testing, or any other environmental conditions. In the area of software, it is cost prohibitive to obtain and test N identical softwares independently in order to obtain a value for the parameter λ. We may also need N identical computing machines to ensure fairness or non-bias.

The second technique, MTTR, has the obvious flaw that the time taken to repair is dependent on the competence of the maintenance team, and therefore cannot be operationally predictable or statistically standardized. Moreover, time to repair depends on the nature of the bug. Since this is universally unpredictable, such an MTTR measure is relative to a given bug; we may be tempted to compute different MTTR for different bugs, and perhaps obtain a mean MTTR (MMTTR). It follows therefore that this method is reliably unreliable. However, there is a way to make this technique realistic: for a well-structured system, compute MTTR for each module (or function) and take the statistical mean. This will cancel out the effects of different bugs due to corresponding modules (or functions).

These anomalies suggest we use the properties of software systems to define their reliability. Two methods have been advocated [46]:

3. Percent up-time: Use the percentage of operational time as the reliability.

4. Number of bugs vs. calendar months: Use the slope of this graph as a rough measure of productivity.

However, (3) will never be known until failure occurs; (4) will take considerable time to obtain, in addition to its dependence on the detectable bugs present.

Against these four techniques, something is obviously deficient—a software goes into production long before a reliability is computed for it. Essentially, we are working with a software whose reliability is unknown, until failure occurs. Another question is how often the software is run. If the frequency (of production) is small, then we need a longer time to failure before computing reliability. The method developed here will correct for this anomaly, as well as be consistent with the philosophy of PET.

We shall develop a simple formula that will give us instant (tentative, if necessary) reliability of a software, that is, before going to production (in fact at design) we already have the reliability factor at hand. Then as production continues, we can obtain actual "failures" and substitute them into our formula to obtain "operating" or "functional" reliability for that software. Moreover, the bugs discovered at test runs may be used in this formula to obtain a functional reliability.

One solution to this may be to count and classify all the possible bugs and compute a simple probability mass function as p_i. However, this does not indicate which of these bugs will surface as failure. The use of functions as the focus of attention seems to be more reliable, since a knowledge of the functions *may* indicate which bugs will be present.

Let a software system consist of N modules. Alternatively, let a module consist of N functions. Let the probability of failure of the i-th module (function) be p_i. Then the reliability of the i-th module is $1 - p_i$. Since the probability of failure for any module (function) is a random event, and since only one bug can cause an *observed* failure, our theoretical reliability will assume that of a mutually exclusive model. Theoretically,

$$r_i = 1 - p_i = 1 - \frac{1}{N} \tag{7.5B}$$

In particular, we shall assume the M chains are independent random variables and compute r_i for each chain. For greater precision, we may choose $N = [\Sigma C_i]$ where C_i = degree of computational complexity of the N_i node (the symbol [ ] indicates "largest integer.") Then, over a length of time, T, the expected value of reliability, R, becomes:

$$\hat{R} = E(R) = \sum_{i=1}^{M} M_i(1 - p_i) \tag{7.5C}$$

Thus, by this approach, we practically tie our measure to the very ingredients of our object—the modules (functions). Notice that $R \neq 0$, and since the software is far from being ideal, $R < 1$ or 100%. Note also that $p_i = F_T(t)$ as introduced at the beginning of this section.

In practice, we may choose $p_i = 0.5$ or 0.8 or any other computed or arbitrary value, depending on the competence of the designer, for all i or compute the p_i's during the testing phase. Thus, before going to production, R is tentatively known.

EXAMPLE 7.5A:

If our software system has 4 modules, we then estimate that given the probability mass distribution: $p_1 = 0.9$; $p_2 = 0.75$; $p_3 = 0.8$; and $p_4 = 0.85$.

$$E(R) = (1 \times 0.9) + (2 \times 0.75) + (3 \times 0.8) + (4 \times 0.85) = 8.2$$

For D_R, degree of reliability, we normalize $\hat{R} = E(R)$ by dividing by ΣM_i to obtain

$$D_R = \frac{E(R)}{\Sigma M_i} \tag{7.5}$$

For the system of this example, $D_R = 0.820$. Clearly, this is a modest low-end estimate, which is better than assuming a very high-end of say 90%–95% for MTBE or MTTR, only to be later found disappointingly exaggerated.

EXAMPLE 7.5B: D_R/Figure 3.3A

Let us compute D_R for the system of Figure 3.3A.

We need only compute for the chains at level 1. Their actual probability for each chain is computable in detail from their refined subcomplexes—a reflection on D_c and D_s. Naturally, the more refinements out of a chain, the more the likelihood that reliability is degraded. A serious effort to compute these probabilities is recommended. For instance, it is clear that simpler functions like 0,1 and 0,5, having to do with OPENing and CLOSing, respectively, of files, will have a high probability of satisfactory computation and may thus be assigned $p = 0.95$.

Assuming that the following probabilities have been computed:

$$p_1 = p_5 = 0.95; p_2 = 0.85; p_3 = 0.75; \text{ and } p_4 = 0.90$$

then we obtain by (7.5):

$$D_R = \frac{1 \times 0.95 + (2 \times 0.85) + (3 \times 0.75) + (4 \times 0.90) + (5 \times 0.95)}{15}$$

$$= \frac{13.25}{15} = 0.8833$$

In contrast, the theoretical probability yields:

$$r_1 = 1 - \frac{1}{3} = .667;$$

$$r_2 = 1 - \frac{1}{4} = .750;$$

$$r_3 = 1 - \frac{1}{5} = .800;$$

$$r_4 = 1 - \frac{1}{7} = .857;$$

$$r_5 = 1;$$

therefore

$$D_R = \frac{1(.667) + 2(.75) + 3(.8) + 4(.857) + 5(1)}{15}$$

$$= \frac{12.995}{15} = 0.8663$$

In conclusion, the formulae of Part Two for measuring the attributes of software, having been based on the properties of the software structure, are easy to compute. They are the direct consequences of the principles of structured programming. Besides negating the arguments against structured programming, these formulae vindicate the new aphorism that programming is the intellectual management of complexity and that the practical philosophy toward this end is structured programming.

EXERCISES

7.1 Construct an alternative definition of degree of software complexity. What are the significant factors for determining software complexity? Which ones are critical?

Compute D_c for figures 5.3.2B and 5.4.2B.

7.2 Do same for software Stucturedness.

Compute D_s for figures 5.3.2B and 5.4.2B.

7.3 Construct an alternative definition of requirements definition. Compare it with that presented here.

7.4 Construct an alternative definition of software reliability. Criticize the approach in this text.

7.5 Construct an alternative definition of software correctness. What are the differences between software correctness and software reliability? Discuss the two methods of demonstrating software correctness.

BIBLIOGRAPHY

1. Aaron, D. "The Superprogrammer Project." *NATO Conference on Software Engineering Techniques*, 1969; Buxton and Randell, editors, Van Nostrand Reinhold, New York.

2. Baker, F.T. "Chief Programmer Team Management of Production Programming." *IBM Systems Journal*; Vol. 11, #1 (1972); pp. 56-73.

3. Baker, F.T. and Mills, H.D. "Chief Programmer Teams." *Datamation*, Vol. 19, #12 (1973); pp. 58-61.

4. Bobrow, L.S. and Arbib, M.A. *Discrete Mathematics; Applied Algebra for Computer and Information Science.* Hemisphere Publishing Corporation, Washington, London, 1974.

5. Böhm, C. and Jacopini, G. "Flow Diagrams, Turing Machines and Languages With Only Two Formation Rules." *Communications of the ACM:* Vol. 9, #5 (1966); pp. 366-71.

6. Boolos, George and Jeffrey, Richard. *Computability and Logic.* Cambridge University Press, London, New York, 1974.

7. Buxton, J.N. and Randell, B., editors. NATO *Conference on Software Engineering Techniques*, Kynoch Press, Birmingham, 1969.

8. Chapin, Ned. "Semi-Code in Design and Maintenance." *Computers and People:* Vol. 27. #6 (1978); pp. 17-27.

9. Chapin, Ned. "New Formats for Flowcharting." *Software Practice and Experience*; Vol. 4, #4 (1974); pp. 341-57.

10. Chapin, Ned and Denniston, Susan P. "Characteristics of a Structured Program." *SIGPLAN Notices*; Vol. 13, #5 (1978); pp. 36-45.

11. Chapin, Ned. "Economic Evaluation in Systems Analysis and Design." *System Analysis and Design: A foundation for the 1980's.* Elsevier North Holland, William W. Cotterman, et al. editors, 1981; pp. 76-90.

12. Dahl, O. -J., Dijkstra, E.W., and Hoare, C.A. *Structured Programming.* Academic Press, New York, 1972, 7th Printing, 1975.

13. De Marco, Tom. *Concise Notes in Software Engineering.* Yourdon Press, Inc. New York, 1979.

14. Denning, Peter. "A Hard Look at Structured Programming." *Infotech State of the Art Reports on Structured Programming;* Infotech International, Ltd. Maidenhead, England, 1976.

15. Denning, Peter, *Concise Survey of Computer Methods.* Petrocelli Books, Princeton, N.J., 1974.

16. Dijkstra, E.W. "Structured Programming." NATO *Conference on Software Engineering Concepts and Techniques,* 1969; Buxton, J.N. and Randell, B., editors, Van Nostrand Reinhold, New York, 1969.

17. Dijkstra, E.W. "The Humble Programmer." *Communications of the ACM:* Vol. 15, #10 (1972); pp. 856–66.

18. Dijkstra, E.W. "Programming Considered As a Human Activity." *Proceedings of the 1965 IFIP Congress.* Amsterdam, The Netherlands; North-Holland Publ. Co.; 1965; pp. 213–17.

19. Dijkstra, E.W. "Go To Statement Considered Harmful." *Communications of the ACM:* Vol. 11. #3 (1968), pp. 147–48.

20. Donaldson, J.R. "Structured Programming." *Datamation,* Vol. 19, #2 (1973); pp. 52–54.

21. Endres, A.B. "An Analysis of Errors and their Causes in System Programs." *IEEE Transactions on Software Engineering;* Vol. SE-1, #2 (June, 1975); pp. 140–49.

22. Freedman, Daniel and Weinberg, M. *Ethnotech Review.* Ethnotech, Inc., Lincoln, NE. 1977, 1979.

23. Gane, Chris and Sarson, Trish. *Structured Systems Analysis.* Prentice-Hall, Inc., Englewood Cliffs, N.J. 1979.

24. Glass, Robert L. *Software Soliloquies.* Computing Trends, Seattle, Wa., 1981.

25. Goodenough, John B. and Gerhart, Susan L. "Toward a Theory of Test Data Selection." *IEEE Transactions on Software Engineering;* Vol. SE-1, #2, (1975); pp. 156–73.

26. Hiller, Frederick S. and Liberman, Gerald J. *Introduction to Operations Research.* Holden-Day, San Francisco, CA., 6th printing; 1970.

27. Hoar, C.A.R. "Prospects for a Better Programming Language." C. Boon, editor; *High-Level Languages; Infotech State of the Art Report,* Vol. 7 (1972); pp. 327–43.

28. Hurley, Richard B. *Decision Tables in Software Engineering.* Van Nostrand Reinhold, New York, 1982.

29. Jackson, Michael. *Principles of Program Design.* Academic Press, New York, N.Y., 1975.

30. Kasner, Edward and Newman, James. *Mathematics and the Imagination.* Simon and Shuster, New York, 5th paperback printing; 1967.

31. Knuth, Donald. "Structured Programming With GOTO Statements." *Current Trends in Programming Methodology.* Raymond T. Weh, editor. Vol. 1, 1977; Prentice-Hall, Englewood Cliffs, N.J.

32. Landin, Peter. "The Next 700 Programming Languages." *Communications of the ACM:* Vol. 9, #3 (1966); pp. 157-66.

33. McCraken, D. "Revolution in Programming: An Overview." *Datamation,* Vol. 19, #12 (1973); pp. 50-52.

34. McGowan, Clement L. and Kelly, John R. *Top-down Structured Programming Techniques.* Petrocelli/Charter, Princeton, Englewood Cliffs, N.J.

35. Miller, E.F. and Lindamood, G.E. "Structured Programming: Top-Down Approach." *Datamation,* Vol. 19, #12 (1973); pp. 55-57.

36. Mills, H.D. "Mathematical Foundations of Structured Programming." IBM Federal Systems Division, Report #SC 72-6012, Gaithersburg, Md., 1972.

37. Mills, H.D. "The New Math of Computer Programming." *Communications of the ACM:* Vol. 18, #1 (1975); pp. 43-48.

38. Mills, H.D. "How to Write Correct Programs and Know It." *Proceedings of the International Conference on Reliable Software,* Los Angeles, Calif.; April 1975. (*SIGPLAN Notices;* Vol. 10, #6 (1975); pp. 363-70.)

39. Myers, Glenford J. *Composite/Structured Design.* Van Nostrand Reinhold Company, New York, 1978.

40. Naur, Peter. "GoTo Statements and Good ALGOL Style." *BIT,* Vol. 3 (1963); pp. 204-208.

41. Parnas, D.L. "On the Criteria To Be Used in Decomposing Systems Into Modules." *Communications of the ACM:* Vol. 5, #12 (1972); pp. 1053-58.

42. Parzen, Emmanuel. *Modern Probability Theory and its Applications.* John Wiley & Sons, New York, 1960.

43. Ramamorthy, C.V. and Ho, Sill-bun F. "Testing Large Software with Automated Software Evaluation Systems." *IEEE Transactions on Software Engineering.* Vol. SE-1, #1 (1975); pp. 46-58.

44. Shore, D.V. "Improved Organization for Procedural Languages." System Development Corporation, Technical Memo TM 3086/002/00 (Santa Monica, Calif.; Sept. 8, 1966).

45. Strachey, C. "Varieties of Programming Language." *High Level Languages, Infotech State of the Art Report,* C. Boon, editor; Vol. #7 (1972).

46. van Tassel, Dennie. *Program Styles, Design, Efficiency, Debugging, and Testing.* Prentice-Hall, Englewood Cliffs, N.J., 2nd ed., 1978.

47. Teichroew, D. and Hershey, E.W., III. "PSL/PSA: A Computer-Aided Technique for Structured Documentation and Analysis of Information Systems." *IEEE Transactions on Software Engineering;* Vol. SE-3 1977; pp. 41–48.

48. Warnier, J.D. *Logical Construction of Programs,* trans. by B.M. Flanagan. Van Nostrand Reinhold Company, New York, 1974.

49. Wirth, Nicklaus. "Program Development by Stepwise Refinement." *Communications of the ACM:* Vol. 14, #4 1971; pp. 221–27.

50. Wirth, Nicklaus. "On the Composition of Well-Structured Programs." *Computing Surveys;* Vol. 6, #4 1974; pp. 247–49.

51. Wulf, W.A. "Programming Without the GOTO." *Proceedings of the 1971 IFIP Congress;* Vol. 1; Amsterdam, North-Holland Publ. Co.; The Netherlands; 1972; pp. 408–13.

52. Yourdon, Edward N. *Techniques of Program Structure and Design.* Prentice-Hall, Englewood Cliffs, N.J., 1975.

53. Yourdon, Edward N. and Constantine, Larry. *Structured Design.* Prentice-Hall, Englewood Cliffs, N.J. 1975.

54. Yourdon, Edward N., editor. *Classics in Software Engineering.* Yourdon Press, New York, 1979.

55. Zwass, Vladimir. *Introduction to Computer Science.* Harper & Row, Publishers, New York, 1981.

56. _____. *Structured Programming. Infotech State of the Art Report.* Infotech International Ltd., Maidenhead, England, 1976.

INDEX